T0174686

The Nile Hydroclimatology:
Impact of the Sudd wetland

DISSERTATION

Submitted in fulfilment of the requirements of
the Board for Doctorates of Delft University of Technology
and of the Academic Board of the UNESCO-IHE Institute for Water Education
for the Degree of DOCTOR
to be defended in public
on Thursday, 30 June 2005 at 10:30 hours
in Delft, the Netherlands

by

Yasir Abbas MOHAMED

born in Sennar, Sudan
Master of Science (with distinction), UNESCO-IHE

This dissertation has been approved by the promotors
Prof. dr. ir. H.H.G. Savenije   TU Delft/UNESCO-IHE Delft, The Netherlands
Prof. dr. W.G.M. Bastiaanssen   ITC, WaterWatch, The Netherlands

Members of the Awarding Committee:
Chairman   Rector Magnificus TU Delft, The Netherlands
Co-chairman   Director UNESCO-IHE, The Netherlands
Prof. dr. ir. H.H.G. Savenije   TU Delft/UNESCO-IHE Delft, The Netherlands, promotor
Prof. dr. W.G.M. Bastiaanssen   ITC, WaterWatch, The Netherlands, promotor
Prof. dr. ir. N.C. van de Giesen   TU Delft, The Netherlands
Prof. dr. S. Uhlenbrook   UNESCO-IHE, The Netherlands
Prof. dr. F. Rijsberman   IWMI, Sri Lanka
Prof. dr. B.J.J. van den Hurk   KNMI, The Netherlands
Prof. dr. ir. C. van den Akker   TU Delft, The Netherlands, reservelid

Published by A.A. Balkema Publishers, a member of Taylor & Francis Group plc.
                www.balkema.nl and www.tandf.co.uk

ISBN 0 415 38483 4 (Taylor & Francis Group)

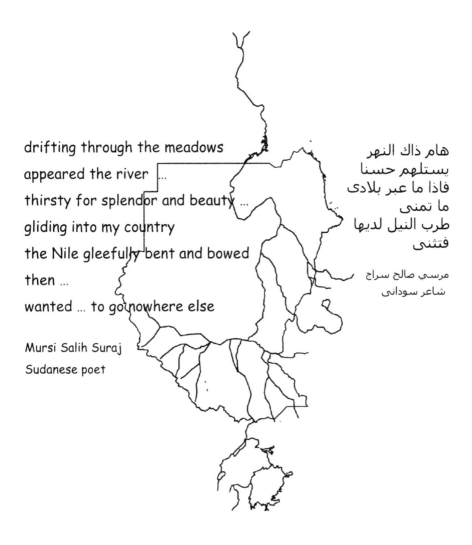

drifting through the meadows

appeared the river ...

thirsty for splendor and beauty ...

gliding into my country

the Nile gleefully bent and bowed

then ...

wanted ... to go nowhere else

Mursi Salih Suraj
Sudanese poet

هام ذاك النهر
يستلهم حسنا
فاذا ما عبر بلادى
ما تمنى
طرب النيل لديها
فتشى

مرسي صالح سراج
شاعر سوداني

الى الشعب السودانى
to   The Sudanese People

Abstract

Climatology and hydrology are increasingly merged into one discipline – in particular at the river basin level – due to the coupled nature of the land surface-atmosphere interactions. Land surface processes – e.g., through evaporation – may affect atmospheric moisture transport, not only locally but also at continental scales. Atmospheric moisture feedback is a significant process that enhances precipitation mechanisms. The atmospheric component of the water cycle should be considered when aiming at a complete analysis of the water resources at river basin scale.

The Nile Basin is characterized by fast declining water resources available per capita in the downstream areas, and large evaporation from the Upper Nile swamps. Since many years, there have been plans to augment the Nile flow by reducing evaporation from theses wetlands through river short cut channels (e.g., the Jonglei canal). However, the question is whether the evaporation from the upland wetlands contributes also to the regional precipitation through moisture recycling? The key question is how much is the net gain in the regional water cycle if the wetlands are drained?

The objective of this research is to determine and analyze the amount of moisture recycling in the Nile basin, so as to quantify the impact of draining the Sudd wetland on local and regional hydroclimatology. Regional climate modeling has been the basic tool of the investigation.

First – as data on the Upper Nile swamps is very scanty – remote sensing techniques have been employed to estimate the actual evaporation from these vast swamps, being a fundamental boundary conditions for the hydrological and climate models. The SEBAL algorithm has been used to compute the actual evaporation from NOAA-AVHRR images. More than 115 satellite images over an area of 1000 km x 1000 km (located on the Upper Nile) have been processed to prepare monthly evaporation and biophysical properties maps for 3 years of different hydrometeorological conditions (1995, 1999 and 2000). The evaporation results were verified against water balance calculations in 3 of the wetlands: Sudd, Bahr el Ghazal and the Sobat. A close resemblance was obtained for the Sudd (2% error), and the Sobat (5% error), while the balance lacks closure for the Ghazal basin (27% error) due to inadequately gauged flow from the upper catchments. The evaporation rate of the Sudd wetland is 20% less, and the average area occupied by the wetland is 70% larger, than if it is assumed to resemble an open water body, an assumption that persisted for decades in hydrological studies of the Sudd.

Further in-depth investigation has been made of the biophysical properties derived from satellite images (surface albedo, surface emissivity, leaf area index, roughness height, aerodynamic resistance, surface resistance) to interpret the temporal variability of evaporation over the wetlands. It is demonstrated that the variation of the atmospheric demand in combination with the inter-annual fluctuation of the ground water table – due to a pronounced seasonality in rainfall and inundation by the Nile floods – results into a quasi-constant evaporation rate over the Sudd through-out the year.

The spatially distributed evaporation data from remote sensing over the Sudd, forms one of the key data sets for validating a Regional Atmospheric Climate MOdel (RACMO) enclosing the Nile basin. The model boundary is between 12°S and 36°N and between 10°E and 54.4°E, with 50 km horizontal resolution. It is forced by ERA-40 data (ECMWF Re-analysis 1957-2001). The model has been adjusted to fit the complex hydrology of the Upper Nile wetlands. The Nile runoff generated over the upstream catchments has been routed to feed the wetland system, which subsequently evaporates. Modest changes have been incorporated into the parameterization of the orography (filtering), solar radiation (aerosols), runoff (drainage coefficient) and soil moisture (layer depth), which produced improved model results. Observational data sets were used to evaluate the model results including: radiation, precipitation, runoff and evaporation data. The model provides an insight not only into the temporal evolution of the hydroclimatological parameters of the region, but also the land surface-climate interactions and embedded feedbacks. The model is used to describe the regional water cycle of the Nile basin in terms of: atmospheric fluxes, land surface fluxes and land surface-climate feedbacks. The monthly moisture recycling ratio (i.e., locally generated precipitation over total precipitation) over the Nile varies between 8 and 14%, with an annual mean of 11%, which implies that 89% of the Nile basin precipitation originates from outside the basin's physical boundaries. The monthly precipitation efficiency varies between 12 and 53%, and the annual mean is 28%. The mean annual moisture recycling ratio in the Nile basin is 11%, which is somewhat, lower than in the Amazon, but of the same order of magnitude as in the Mississippi basin.

The impact of the Sudd wetland on the Nile hydroclimatology has been studied by comparing two regional climate model scenarios: the present climatology, and a drained Sudd scenario. The results indicate that in general the Sudd has negligible impact on the regional hydrological budget terms owing to the relatively small area covered by the wetland. The amount of moisture recycling by the swamps is small compared to the atmospheric fluxes over the region. On the other hand, the runoff gain, in this extreme scenario, would be substantial: about 36 Gm^3/yr, which is about half the natural flow at Aswan. However, the impact on the microclimate is large. The near-surface relative humidity would drop by 30 to 40% during the dry season, and the temperature would rise by 4 to 6 ° over the region that is currently occupied by the Sudd wetlands. The impact during the wet season (June to September) is comparatively small.

Acknowledgement

I was lucky to accomplish this study, getting the opportunity of supervision from three supervisors having different backgrounds. I'm grateful to my promoter Prof. Savenije for the constructive discussions and guidance, not only during this study, but all the way during the last 15 years. From him I learned what "integrated water resources management" is, and why hydrological and climatological process should be coupled if we want to make a complete analysis and understanding of a regional scale water resources system.

I would like to express my gratitude to Prof. Bastiaanssen (co-promoter) who gave me the confidence to learn and use remote sensing techniques for hydrological applications, for this study, and for many other applications relevant and very much required in my country.

I'm very grateful to Prof. Bart Van den Hurk (advisor of the study) for his patience to teach a hydrologist about the fundamentals of a climate model and how it needs to be properly operated. I would like to thank him for all the assistance he provided on the climate modeling part of the study.

This study would not have been possible without the financial support of the International Water Management Institute (IWMI), the International Institute for Geo-information and Earth Observations (ITC), and the Netherlands NFP program. So I would like to express my sincere gratefulness to these three organizations. I should also thank the hosting organizations and their staff who provided all help possible during the course of the study: UNESCO-IHE in Delft, ITC in Enschede, the Royal Netherlands Meteorological Institute (KNMI) in De Bilt, and the Sudanese Ministry of Irrigation in Khartoum.

I take this opportunity to thank two (forgotten departments): the computer department of the Sudan meteorological cooperation, and the Nile Water department of the Ministry of Irrigation. They generously provide all the required data for the case study.

Last, but not least I would like to thank my wife "Amany" of not only being a mother and father of 4 children during my absence, but also for the continuous help in getting the data, keying it into the computer and e-mailing it to the Netherlands. Without her assistance it would have been very hard to obtain all these masses of data.

List of symbols

Symbol	Interpretation	Unit
A	area	m^2
A_w	area of water surface	m^2
A_{unsat}	area of unsaturated zone	m^2
B	Bowen ratio	-
c_p	air specific heat at constant air pressure	J/kg/K
C_R	runoff coefficient	-
d	displacement height	M
E	evaporation rate	$Kg/m^2/s$
E_0	reference crop evaporation rate	$Kg/m^2/s$
E_a	actual evaporation rate	$Kg/m^2/s$
E_p	potential evaporation rate	$Kg/m^2/s$
E_w	open water evaporation rate	$Kg/m^2/s$
E_{wet}	wet season evaporation rate	$Kg/m^2/s$
e_a	water vapor pressure	kPa
e_s	saturated water vapor pressure	kPa
G	infiltration rate to ground water	$Kg/m^2/s$
G_0	surface soil heat flux	W/m^2
H	sensible heat flux	W/m^2
h	plant height	m
I_{NDV}	normalized difference vegetation index	-
K_c	crop coefficient	-
k	von Karman's constant	-
L	domain length	m
m	moistening efficiency	-
p	precipitation efficiency	-
P	precipitation rate	$Kg/m^2/s$
P_a	precipitation rate from advected sources	$Kg/m^2/s$
P_l	precipitation rate from local sources	$Kg/m^2/s$
Q	atmospheric moisture flux	$Kg/m^2/s$
Q_a	atmospheric moisture flux of oceanic origin	$Kg/m^2/s$
Q_{in}	incoming atmospheric moisture flux	$Kg/m^2/s$
Q_l	atmospheric moisture flux of recycled origin	$Kg/m^2/s$
Q_{out}	outgoing atmospheric moisture flux	$Kg/m^2/s$
R	runoff rate	$Kg/m^2/s$
R_{in}	incoming runoff rate	$Kg/m^2/s$
R_{out}	outgoing runoff rate	$Kg/m^2/s$
R_n	net radiation flux at land surface	W/m2
R_{nl}	net long wave radiation flux at land surface	W/m^2
R_{sd}	incoming short wave radiation flux at land surface	W/m^2
R_{ld}	incoming long wave radiation flux at land surface	W/m^2
r	soil moisture recharge rate	$Kg/m^2/s$
r_0	surface reflectance (albedo)	-
r_a	aerodynamic resistance to water transport	s/m
r_c	canopy resistance to water transport	s/m
r_s	bulk surface resistance to water transport	s/m

RH	relative humidity	-
S	storage volume	m^3
S_{unsat}	storage volume in the unsaturated zone	m^3
S_w	storage volume of surface water	m^3
T	temperature	K
T_0	land surface temperature	K
T_{01}	land surface temperature over horizontal plain	K
T_a	air temperature	K
u_z	wind speed	m/s
u_{100}	wind speed at the blending height	m/s
W	atmospheric moisture content	Kg/m^2
z	height	m
z_{0m}	roughness height for momentum transfer	m
z_{0h}	roughness height for heat transfer	m
ρ_a	air density	kg/m^3
Δ	slope of the saturated vapor pressure curve	$kPa/C°$
γ	psychrometric constant	$kPa/C°$
ζ	feedback ratio	-
ε_0	thermal infrared emissivity	-
λ	latent heat of vaporization	J/Kg
λE	latent heat flux density	W/m^2
β	moisture recycling ratio	-
Λ	evaporative fraction	-
α	loss coefficient	-
θ	soil moisture	m^3/m^3
θ_{sat}	soil moisture at saturation	m^3/m^3

List of Acronyms

AVHRR	Advanced Very High Resolution Radiometer
BSRN	Baseline Surface Radiation Network
CATCH	Coupling of the Tropical Atmosphere and Hydrological Cycle
CMAP	Climate Prediction Center Merged Analysis of Precipitation
CRLE	Complementary Relationship Lake Evaporation
CTL	Control scenario
DEM	Digital Elevation Model
DRA	Drained scenario
ECMWF	European Center for Medium Range Whether Forecast
EFEDA	Echival First Field Experiment in Desertification Threatened Area
ERS	European Remote Sensing
ERA-40	ECMWF reanalysis data
EURASIA	European part of the former Soviet Union
FEWS	Famine Early Warning System
FIFE	First ISLSCP Field Experiment
GCM	General circulation model
GEWEX	Global Energy and Water Cycle Experiment
GFDL	Geophysical Fluid Dynamics Laboratory

GLCC	Global land Coverage Characteristics
GPCC	Global Precipitation Climatology Centre
GWT	Ground Water Table
HAPEX	Hydrological and Atmospheric Pilot Experiments
HIRLAM	HIgh Resolution Limited Area Model
IPCC	Intergovernmental Panel on Climate Change
ITCZ	Inter-Tropical Convergence Zone
JIT	Jonglei Investigation Team
KNMI	The Royal Netherlands Meteorological Institute
LAC	Local Area Coverage
MIRA	Microwave Infra Red Algorithm
NCEP	National Centers for Environmental Prediction
NOAA	National Oceanic and Atmospheric Administration
NVAP	NASA Water Vapor Project
PBL	Planetary Boundary Layer
P-M	Penman-Monteith equation
PJTC	Permanent Joint Technical Commission
RACMO	Regional Atmospheric Climate MOdel
RCM	Regional Climate Model
SEBAL	Surface Energy Balance Algorithm for Land
SAA	Satellite Active Archive
SVAT	Soil Vegetation Atmosphere Transfer Parameters
TESSEL	Tiled ECMWF Scheme for Surface Exchanges over Land
WRMC	World Radiation Monitoring Center

Table of content

1. Introduction

1.1 General

The classical hydrological approach is to confine a regional water resources system to only the land hydrological cycle: precipitation P, evaporation E, infiltration G, and runoff R, assuming fixed atmospheric conditions (see Fig. 1.1). However, measurements and modeling studies in many river basins (e.g., Amazon, Mississippi) have confirmed a significant contribution of local evaporation to the atmospheric moisture, and hence to the regional precipitation (Eltahir and Bras, 1996; Bosilovich and Schubert, 2001). Therefore, the atmospheric component of moisture recycling should also be accounted for when aiming at a complete and thorough analysis of the regional scale water resources. Fig. 1.1 shows the components of the water cycle in a region of domain length L. Q_{in} and Q_{out} are the incoming and outgoing atmospheric moisture fluxes, and W is the atmospheric moisture content.

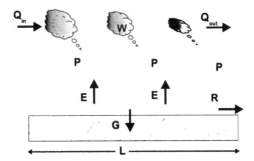

Fig. 1.1: The components of the water cycle in domain L

During the last decades, there has been intensive research on land surface-atmosphere interactions and moisture recycling studies, from various perspectives: climatology, hydrology and plant physiology. Though a good understanding is obtained on the importance of the regional evaporation to sustain rainfall, results vary substantially between different studies, and no concrete findings are available on the land surface-atmosphere interactions at specific temporal and spatial scales. On the other hand, these findings, which are well known and accepted within the climate community, are not yet fully exploited within the water resources community.

1.2 Problem definition

The main problem of the Nile water resources system is characterized by increasing water demands on the downstream areas, and high evaporation from wetlands on the upstream catchments. About half of the Nile flow evaporated when passing through the Sudd wetland. The huge evaporation from Upper Nile wetland has, through the ages, attracted planners searching for additional river flows to build short cut channels to prevent spilling over the swamps (e.g., the uncompleted Jonglei canal).

The central question of this research is: what are the implications of draining the Sudd wetland on the Nile hydroclimatology? Stated differently, does the moisture evaporated from the Sudd contribute as precipitation elsewhere in the region? To resolve this question we have to address a number of sub-questions. First, how much is the actual evaporation from these wetlands (characterized by scarce ground data)? Secondly, what are the land surface-atmosphere interactions and feedback mechanisms that exist over the Nile region? Thirdly, if the complete Sudd is drained, what is the impact on the local and regional climate, in particular the Nile Water cycle, and how much gain of the Nile water can be obtained?

1.3 Research Objective

The main objective of this study is to assess the amount of moisture recycling in the Nile basin, and hence to quantify the impact of draining the Sudd wetland on local and regional hydroclimatology. The specific objectives to reach this conclusion are:

1. Define as accurately as possible the spatial and temporal variability of evaporation, over the Sudd wetland using remote sensing techniques. Verify the derived evaporation estimates.
2. Assess the ability of basin-averaged moisture recycling formulae to quantify impact of land use change on climate.
3. Develop and validate a regional climate model enclosing the Nile Basin to:
    a. Understand the land surface-atmosphere interaction processes in the region, and
    b. Quantify the impact of draining the Sudd wetland on the regional hydroclimatology, including the implication on the Nile flow.

1.4 Outline of the thesis

Can the impact of evaporation on climate be quantified by a basin average moisture recycling formula (mass balance only) – which would save a lot of computation effort – or should the whole land surface-atmosphere feedback loops be accounted for? This question has been addressed in Chapter 2, through an intensive literature review on moisture recycling mechanisms, methods of computation and case studies from the literature. Significance and limitations of the available methods are addressed.

The description of the main features of Nile climate, hydrology and water resources is given briefly in Chapter 3, as a background to evaluate the climate model results. This chapter also presents a brief description of the Sudd wetland.

The actual evaporation from the Sudd wetland has been a debatable subject for several decades, and different figures emerge concerning the size of the wetland and the evaporation rate. Since the Sudd evaporation is a key parameter in the main research question, a through investigation has been carried out to define the spatial and temporal variability of the Sudd evaporation using remote sensing techniques. The spatial distribution is presented in Chapter 4, and the temporal distribution in Chapter 5.

The application of the Regional Atmospheric Climate Model (RACMO) to the Nile present climatology is presented in Chapter 6 (control run). This chapter presents and discusses model results versus observations with particular emphasis on the hydrology of the upstream wetlands.

Chapter 7 presents the application of the climate model to a modified land surface scenario (drained Sudd run). Comparison of the two scenarios to evaluate the impact of the Sudd on local climate and basin hydroclimatology is presented and discussed. The implication on the Nile water flow is quantified.

An overall summary and conclusion of the study is given in Chapter 8. The strong and week points of the numerical experiment are discussed. Further studies on the limitations encountered have been suggested for future research and measurements.

2. How important is land evaporation to sustain regional precipitation? A literature review on moisture recycling.[1]

2.1 Introduction

Land use affects the atmospheric moisture conditions, not only locally but also at continental scales. Hydrologists and water managers are becoming more aware of the need to understand the atmospheric part of the regional water cycle. There is considerable experience present among scientists and water managers: some anecdotal, some empirical, some as a result of modeling. There is, however, neither a consensus nor a full picture on the relative importance of the land surface-climate moisture feedback at different temporal and spatial scales.

The components of the water cycle in a region include horizontal and vertical exchanges of moisture (see Fig. 1.1). The horizontal (lateral) moisture movements include atmospheric advection, stream flow of surface water resources (also through canals) and groundwater movement in aquifers. The vertical exchanges at the land surface interface are governed by precipitation and evaporation. The vertical exchange between the land surface and the groundwater system occurs through the mechanisms of infiltration, percolation, capillary rise and artificial extractions. The relation between these components is highly non-linear and complicated by the different spatial and temporal scales of the components.

Moisture recycling is defined as the part of the evaporated water from a given area that contributes to the precipitation over the same area. It is a primary process in the land surface-climate feedbacks of a region, and hence, has direct implications on the regional land and water resources management. In climate modeling land-atmosphere interaction is known to be crucial process in the hydrological cycle.

The classical hydrological approach is to confine a regional water resources system to only the land hydrological cycle (precipitation, evaporation, infiltration, and runoff) assuming fixed atmospheric conditions at a given observation height close to the surface. The response of the atmospheric water component to evaporation and heat releases is usually not described, nor are the feedback mechanisms on the evaporative land surface conditions. Rainfall in West Africa occurs until approximately 2000 km from the coast, but it would only reach 500 km inland if there were no moisture feedback at all (Savenije, 1995). Numerous measurements and studies over the Amazon and the Mississippi basins confirm a significant contribution of local evaporation to the atmospheric moisture, and hence to the

[1] Based on: Mohamed, Y.A., H.H.G. Savenije, W.G.M. Bastiaanssen, B.J.J.M. van den Hurk, 2004. Moisture Recycling: How important is land evaporation to sustain regional precipitation?, submitted to the J. of Climate Change (Special Issue on the Dialogue on Water and Climate)

regional rainfall (Eltahir and Bras, 1996; Trenberth, 1999; Bosilovich and Schubert, 2001). Although moisture recycling studies usually focus on return flows from river diversions and recycling percolation water through groundwater extraction or through the reuse of drainage water, the atmospheric component of moisture recycling must also be accounted for when aiming at a complete and thorough analysis of the regional scale water resources.

There has been intensive research on moisture recycling during the last decades using different techniques, varying from simple atmospheric water balance analysis (e.g., Budyko, 1974; Brubaker et al. 1993; Savenije, 1995), to complicated modeling of the land surface-climate interactions (e.g., Koster et al., 1986; Zheng and Eltahir, 1998; Bosilovich and Schubert, 2002). Though a good understanding is obtained on the importance of the regional evaporation to sustain rainfall, results vary substantially between different studies, and no concrete findings are available on moisture recycling at specific temporal and spatial scales. On the other hand, these findings, which are well known and accepted within the climate community, are not yet fully exploited within the water resources community.

This chapter aims at reviewing the state of the art of the research on moisture recycling, and highlights the worldwide experience on the subject through a review of selected case studies, their results and the associated limitations. Section 2.2 gives a description of the mechanism of moisture recycling and the effect of soil moisture on the climate feedbacks. Section 2.3 presents a review of the most widely used methods for computation of moisture recycling, followed in section 2.4 by case studies from selected river basins. The experience learned from the international electronic discussion forum on moisture recycling held during September to December 2002 is reported and discussed in section 2.5. Finally, discussion and conclusion on the state of the art on moisture recycling research is given in section 2.6.

2.2 Moisture recycling mechanism

The sources of precipitation in a given region are: the moisture fluxes advected into the region by moving air masses Q_{in} and the moisture flux supplied by evaporation from within the region itself E (see Fig. 1.1). Therefore, the total precipitation P in a domain is composed of two components: P_l from a local source (local evaporation) and P_a from an advected source (oceanic or upstream continental evaporation). The relative contribution of oceanic and continental moisture to the precipitation in a region was subject of discussion for many decades (a historical review is given in Brubaker et al., 1993; Brude and Zangvil, 2001). From the precipitated water in a region, part E evaporates from land surface (including interception, evaporation from open water, from bare soil, and transpiration from vegetation). Part R drains at the outlet of the basin as runoff. Part G is the storage in the subsoil and overland. A major part of G will feed the streams and rivers and are responsible for the annual base flow. Moisture recycling is defined as the part of the precipitated water which evaporated from a given area that contributes to the precipitation over the same area, also referred to as locally derived precipitation. The recycling ratio β is computed as the ratio of P_l/P. Thus it is equal to unity for the whole globe, and zero for a point location.

The moisture recycling process characterizes a non-linear relationship between the regional evaporation, moisture transport, and precipitation. Land surface evaporation depends essentially on 3 factors:

- Surface radiative energy supply.
- Soil moisture conditions.
- State conditions of the atmospheric boundary layer.

All three categories of properties vary with land use. Therefore, land use has a significant impact on the moisture feedback processes through the mechanism of evaporation. A wetland surface has a lower surface albedo (Bowers and Hanks, 1965) and lower surface temperature (Idso et al., 1969). This results in higher absorption of short wave radiation and a lower emittance of long wave radiation, which yields a substantially higher net radiation. As a consequence, the total available heat energy (latent and sensible heat flux together) to the atmosphere is greater than for dry land. For instance, measurements over a semi-arid region in New Mexico by Small and Kurc (2001) showed that adding 5% to the volumetric soil water content yields an increase of 50 W/m^2 of the available energy to the atmosphere. Castelli et al. (1999) confirmed that the evaporative fraction of the energy balance (latent heat flux divided by net available energy) is to a large extent controlled by soil moisture. The correlation of soil moisture with net radiation and latent heat flux is confirmed to exist in many earlier large-scale field investigations such as FIFE (e.g., Smith et al 1992), EFEDA (e.g., Bastiaanssen et al., 1997), HAPEX (e.g., Gash et al. 1997), MONSOON (e.g., Kustas, 1990), and SGP (Jackson et al, 1999). Fig. 2.1 shows a schematic diagram of the feedback processes related to change of soil water content.

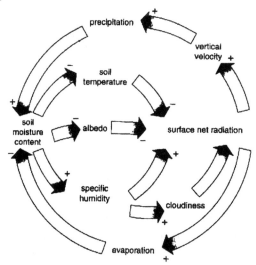

Fig. 2.1: Feedback processes related to a change in soil water content (Rowntree, 1988)

It is known from both observations and numerical experiments, that evaporation has two effects on the atmospheric state conditions:

1.   It enhances the absolute humidity and thus precipitation (e.g., Driedonks, 1982; De Bruin and Holtslag, 1989). Observational data over the Amazon and other regions (see e.g., Eltahir and Bras, 1994; Trenberth, 1999) show a significant contribution of local evaporation to the atmospheric moisture. The relative importance depends upon the amount of the oceanic moisture advected into the region, i.e., local evaporation will contribute greatly to the atmospheric moisture content when the advected moisture is small. Bosilovich and Schubert (2001) computed a smaller recycling ratio of 20% over the central United States during the high flood of 1993, when large amounts of moisture were advected into the region. This ratio rises to more than 60% during the same month of the dry year of 1988, which were associated with smaller amounts of advected moisture.

2.   It affects the overall thermo-dynamic condition of the atmosphere. A higher total energy to the atmosphere leads to larger moist static energy of the boundary layer. Moist static energy plays an important role in the dynamics of the local convective storms, and it strengthens the large-scale monsoon circulation (see e.g., Beljaars et al, 1996; Eltahir, 1998; Schär et al., 1999). It should be mentioned that this particular influence on the thermodynamic of the vertical water column is still a research subject, which undergoes international debates. E.g., using numerical experiments over the Mississippi Giorgi et al. (1996) indicated a negative feedback mechanism of evaporation, that is, a decreased evaporation increases the buoyancy, which dynamically sustains convection, and hence increases the precipitation.

An example of these relationships between soil moisture, radiative properties, net radiation and energy fluxes was recently demonstrated by Mohamed (2004) for the Sudd wetlands over a large part of the Nile basin (see Fig. 2.2) using satellite measurements. The image contains contrasting land surface types: dry land, Savannah, tropical forest and wetlands. The dry land (northern part of the image, and southeastern corner) shows lowest soil moisture ($\sim$ 30% of the saturated soil moisture), lowest net radiation ($\sim$ 250 W/m^2), lowest total heat ($\sim$175 W/m^2), and lowest evaporation ($\sim$ 1.7 mm/day), while the reverse is seen over the Sudd swamps (the triangular area between Juba, Wau and Malakal). The soil moisture in the root zone was computed by an empirical relationship between evaporative fraction and soil moisture proposed by Scott et al. (2003), after testing in Spain, USA, Mexico and Pakistan (see Eq. 4.3). The surface fluxes were calculated using the Surface Energy Balance Algorithm for Land SEBAL model (Bastiaanssen et al., 1998a) in conjunction with NOAA-AVHRR (National Oceanic Atmospheric Administration – Advanced Very High Resolution Radiometer) imagery. Detailed description of SEBAL is given in section 4.2.2.

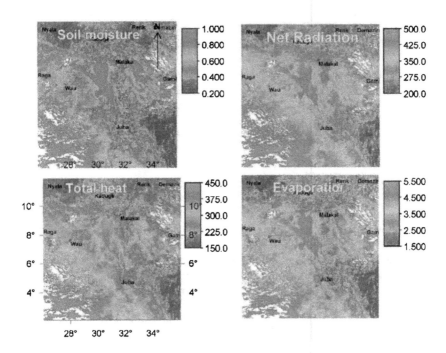

Fig. 2.2: Instantaneous Soil moisture (relative to saturation), net radiation (W/m^2), total heat (W/m^2) and daily evaporation (mm/day) over part of the Nile Basin on 26/01/2000.

2.3 Methods to compute moisture recycling

Almost all researchers agree on the definition of moisture recycling as the ratio of locally generated precipitation to the total precipitation, yet different approaches are used to formulate the recycling method. In general, the methods in use to compute moisture recycling can fall in one of the following two groups:

1. Methods based on the atmospheric moisture balance (e.g., Budyko, 1974; Eltahir and Bras, 1994; Savenije, 1995). These methods are also called the bulk methods. Data used in the calculations can be direct observations, reanalysis data, or pure results from numerical experiments.

2. Methods based on tracing trajectories of the water molecules from the source origin through the atmosphere and then as precipitation (e.g., Koster et al., 1986; Numagati, 1999; Bosilovich and Schubert, 2002). Passive water vapor tracers are introduced into GCM (General Circulation Model) simulations to follow the trajectories of the water molecules within a given atmospheric system.

In the following section, a summary of some of the widely used recycling methods is given, followed by a comparison of the derived results. A review of some of these methods is also given in the literature (e.g., Brude and Zangvil, 2001).

1. The method derived by Budyko (1974), and extended by Brubaker et al. (1993) and Trenberth (1999), defines the recycling ratio β:

$$\beta = \frac{P_l}{P} = \frac{EL}{EL + 2Q_{in}} = \frac{EL}{PL + 2Q} \tag{2.1}$$

where Q is the average flux along a domain of length L, given by $0.5(Q_{in}+Q_{out})$, (see Fig. 1.1), and Q_{in} and Q_{out} are the inward and outward moisture fluxes into the given domain. By rewriting Eq. (2.1) the average horizontal advective flux Q can be expressed as $(Q_{in}-0.5P_aL)$, where P_a is the precipitation component from the advected moisture. The horizontal flux of local origin is $0.5(E-P_l)L$. The basic assumption here is that the atmosphere is well mixed and the change of atmospheric moisture storage is negligible compared to the other terms. The recycling results of course depend on the length of the domain L, which may involve the difficulty of defining the areal extent of the region. To avoid dependence on domain length, Trenberth (1999) computed the recycling ratio for the whole world based on two length scales 500 km and 1000 km, which allows comparison of the different regions. Section 2.4, Table 2.2 gives results of this method among others as applied to the flux data from four basins.

2. Eltahir and Bras (1994) developed a recycling ratio based on conservation of mass of a control volume in the given region. Similar to Budyko's model, two basic assumptions are imbedded: the atmospheric moisture is well mixed, and the rate of change of the atmospheric moisture is negligible on a monthly time scale. The spatially distributed moisture recycling ratio is then defined as:

$$\beta = \frac{P_l}{P} = \frac{Q_l + E}{Q_l + Q_a + E} \tag{2.2}$$

Here Q_{in} is the flux moisture into the control volume, e.g., into a model grid, and it is composed of Q_l (moisture flux from recycled moisture) and Q_a (moisture flux of oceanic or upstream continental origin). For the whole region (e.g., the Amazon basin), $\beta = E/(Q_{in}+E)$. Please note the difference in this case with method one given above (numerator).

3. Koster et al. (1986), Bosilovich and Schubert (2002) used passive tracers of water molecules in GCM simulations to trace the route of evaporated moisture from land surface through the atmosphere and then as precipitation on a different location. Results obtained could show the relative importance of the source regions to regional precipitation. The advantages of this method over the atmospheric water balance (bulky) methods that the tracer method is more accurate but more computationally expensive (Bosilovich and Schubert, 2002).

However, it assumes that advected air masses are unmodified during the advective transfer, which is not realistic when coarse resolution models are used.

4. Savenije (1996a) assumes a Lagrangian movement of the air mass over the Sahel region, where he applies a one dimensional moisture balance of the atmosphere to define moisture recycling. In the rainy season, the net advective moisture along distance L is equivalent to the precipitation component from advection P_a, while local rainfall is equivalent to wet season evaporation E_{wet}. Therefore moisture recycling over the rainy season is given by:

$$\beta = \frac{E_{wet}}{P} = 1 - \alpha \qquad\qquad (2.3)$$

where α is the loss coefficient from the system, which equals the runoff coefficient C_R (runoff divided by precipitation) plus the part of the rainfall which evaporates during the dry season e_d, $\alpha = C_R + e_d$. It is assumed that E_{wet} is completely removed as precipitation, and no evaporated water leaves the region at the downstream end.

5. Schär et al. (1999) assume the feedback mechanism to be composed of two processes: a direct component (recycling), whereby extra precipitation is caused by addition of evaporative moisture into the atmosphere, and an indirect process (amplification), whereby extra precipitation is due to an enhanced precipitation efficiency caused by local evaporation. The recycling formula used over large regions (e.g., France) is:

$$\beta = \frac{E}{E + Q_{in}} \qquad\qquad (2.4)$$

This is the same as the Eltahir and Bras (1996) formula applied to a region, setting the inflow flux at the boundary originated from local evaporation Q_l to zero. Note the difference as compared to the formulae given in Budyko (1974), Brubaker et al. (1993) and Trenberth (1999).

One basic assumption in all the atmospheric water balance methods (bulk methods) is that the atmosphere is assumed to be well mixed both temporarily and spatially (Budyko, 1974). Eltahir and Bras (1994) and Harris et al. (1988) in their investigation of the Planetary Boundary Layer (PBL) over the Amazon, showed that vertical mixing is generally attained in a relatively short time compared to the advective time scale. On the other hand, Bosilovich Schubert (2002) showed that the local moisture and the oceanic moisture are not well mixed and remain vertically stratified, using water vapor tracers over the Central Plain of the United States. In the horizontal direction the assumption of well mixing is even weaker because of the spatial variations of the atmospheric parameters (e.g., variations of temperature, humidity, local conditions) over a wide range of spatial scales. The mixing is directly related to the geometrical structure of the landscape. Shuttleworth (1988) separated two types of landscape heterogeneity: one that exhibits disorganized variability at length scales of 10 km or less (giving no apparent organized response in the atmospheric boundary layer), and one that exhibits variability which is organized at length scales of greater than 10 km (and may give an organized response in the atmosphere such as to alter the effective value of surface properties) respectively.

It should be emphasized that the precipitation recycling ratio as defined above is a diagnostic measure that defines the contribution of local evaporation to local precipitation under given climate conditions (Brubaker et al., 1993; Trenberth, 1999; Eltahir and Brass, 1996). The precipitation recycling ratio has no prognostic value, i.e., it is not a fixed variable for all climate conditions in a region. This is because of the non-linear relationship between precipitation and evaporation in the given climate system of a region. Thus, the quantitative results of moisture recycling studies are to be considered as regional scale characteristic of a certain basin.

2.4 Case Studies

Worldwide, there are many studies aiming at quantifying the regional moisture recycling, and this review does not attempt to be complete. Most of the studies showed increase of precipitation with increasing evaporation. However, there are substantial variations depending on the model used, data source, location and season of the year. Prove of the results with real life observations is restricted by the limitations of data availability and its reliability. As a result it is not unusual to find different recycling ratios for a given region e.g., for the Amazon or the Sahel computed by different researchers. Also different results may be found in the same basin, with the same input data, by using different definitions of moisture recycling. E.g., Bosilovich and Schubert (2001) have computed two different summer time recycling ratios over the Central United States with the same data set: using the method proposed by Brubaker et al. (1993), the recycling ratio is 25%. The recycling ratio increases to 36% when the Eltahir and Bras (1996) method is used.

In the following, a comparison of some of the results in the literature of the annual regional moisture recycling is presented (see Table 2.1). It should be emphasized that most of these results were originally derived on monthly basis showing large seasonal variations, which were smoothed in the annual results.

To compare the results of moisture recycling by the formulae given in section 2.3, we make use of the data on the annual regional water cycle of the Amazon, Mississippi, Nile and the Sahel (see Table 2.2). The data on the Amazon were derived from ECMWF 15-year reanalysis data (Eltahir and Bras, 1996). For the Mississippi they are from Benton et al. (1950). The data for the Nile Basin (between 4°S and 24°N, 23°E and 40°E) are derived from the results of the Nile regional climate model (Chapter 6). For West Africa the data were derived from the model of Savenije (1995). Numbers are yearly totals normalized by the yearly precipitation (index = 100).

Table 2.1: Example of average annual moisture recycling over different regions

Basin	Amazon	Mississippi	West Africa	Eur-asia*	Method and data
Budyko (1974)				11%	Budyko model along a streamline, and observed data of various sources.
Molion (1975)	56%				Based on the ratio of total evaporation to total precipitation in the Amazon basin.
Brubaker et al. (1993)	24%	24 %	31%	11%	Extended 1 D Budyko model, and analyzed observations. (West Africa is the Niger Basin).
Eltahir and Bras (1994)	25% 35%				Using spatially distributed recycling model, and two data sources: the ECMWF* gives (25%), and GFDL* gives 35%.
Savenije (1995)			63%		This is point-recycling ratio in the Sahel, based on atmospheric and terrestrial water balance analysis. It is applicable during rainy season only.
Trenberth (1999)	34%	21%			Using spatial model based on 500 km length scale. Amazon length is 2750 km, and Mississippi is 1800 km. Data from CMAP*, NVAP* and NCEP*.
Bosilovich et al. (2002)	50 %	47%	44%		Using water vapor tracers in a GCM. West Africa is the CATCH* GEWEX* experiment. The recycling ratio here means moisture of continental origin.

*see list of acronyms for explanation of the abbreviations

It is to be noted that there can be substantial spatial variation of the moisture recycling ratio within the basin itself. Based on the spatially distributed calculation of the recycling ratio for the 500 km length scales, Trenberth (1999) calculated the recycling ratio for the Amazon and Mississippi basins to be twice as high as the estimates in this study (see Table 2.1). The recycling ratio estimated by Eltahir and Bras (1994) based on 2.5x2.5° grid calculations (given in Table 2.1) resembles the value of the basin as a whole given in Table 2.2. The annual recycling ratio as computed in Table 2.2 conceals substantial seasonal variations. E.g., Brubaker et al. (1993) computed for the Mississippi a ratio of 15% during the winter season, and up to 34% during the summer months.

Table 2.2: Annual moisture recycling ratio over the Amazon, Mississippi, Nile and the Sahel by different methods.

Basin	Amazon	Mississippi	Nile	West Africa (Sahel)
Input data (annually)				
Q_{in}	141	466	365	
Q_{out}	99	444	351	
P	100 = 1950 mm/yr	100=750 mm/yr	100=557 mm/yr	100=870 mm/yr
E	58	78	86	
R	42	22	14	
E_{wet}	<58	n.a.	33	63
Recycling ratio				
Brubaker et al. (1993)	17%	8%	11%	
Eltahir and Bras (1994)	29%	14%	19%	
Savenije (1996a)*	<58%	n.a.	33%	63%
Schär et al. (1999)	29%	14%	19%	

* wet season is from Jun to Sep

Difficulties and constraints in attaining a unified result of moisture feedback are obvious, and due to several reasons. First, there are assumptions utilized in the methods that may not be fully satisfied in reality, e.g., most of the methods (Budyko, Eltahir, Savenije, Schär) assume a well mixed atmosphere in the region under consideration, implying that advected and evaporated moisture are well mixed. It is well possible that this assumption is not fully satisfied, i.e., there can be variations in temperature, humidity, and orographic effects along the trajectory of atmospheric moisture that prevents optimal mixing. As moisture recycling is directly dependent on the length of the domain, different results are obtained with different definitions of areal extents of the regions. This can be partly overcome through the use of spatially distributed computations, or unified length scales. On the other hand, results of regional moisture recycling based on numerical experiments are limited to the model veracity in simulating the interaction processes. In this respect, temporal and spatial resolution has a direct impact on the final results. E.g., orography may not be well captured by a model resolution of 5°x5° or even 2.5°x2.5°. Modeling results based on monthly data will ignore the diurnal cycle variation (Bosilovich and Schubert, 2001). The land surface-climate coupling schemes imbedded in the GCM vary between the models, and hence will vary the derived results. In their comparison of the coupling strength in 4 GCM's, Koster et al. (2002) showed significant variations among those models. A continuing problem in the determination of moisture recycling and other hydroclimatological parameters as well, is the availability of data needed to compute the fields on which they are based. Error sources arise from both data sampling and analysis techniques. Finally, because of data scarcity in many regions of the world, the reanalysis data used in the recycling studies may also reflect some of the model bias.

Although far more studies and model experiments support positive moisture feedback, there exist some studies concluding opposite results, or at least do not support positive moisture climate feedback. Giorgi et al. (1996) in their numerical experiments over the Central Unites States for the two climatic extremes (1988 drought and 1993 flood) found that the effect of local recycling of evaporated moisture is not important as compared to the large-scale moisture fluxes and synoptic cyclonic activity. It is even concluded that a dry initial soil condition provides increased sensible heat flux, causing greater air buoyancy, enhancing convective systems and hence providing more precipitation (i.e., a negative moisture feedback process). The hypothesis of Eltahir (1989) that an increase of the wetlands area over part of the Nile Basin (Sudd and Bahr el Ghazal swamps) would favor increased rainfall over central Sudan, and also the argument of Eagleson (1986) that the evaporation from the Sudd would surely be felt climatically over a wider region, is argued by Sutcliffe and Parks (1999).

2.5 The international electronic discussion forum on moisture recycling

An international electronic discussion on moisture recycling – as part of the literature review – has been held during the period Sep to Dec 2002, within the framework of the Dialogue on Water and Climate. The forum aims at raising a discussion on moisture recycling issues: its importance to sustain regional rainfall, the available observational and computational experiences in the world, and what should be the direction of the future research to better understand and quantify moisture recycling. Anticipated participants were invited through e-mails. It mainly concerned scientists working in hydro-climatology and hydrometeorology, although others were invited as well. Comments conveyed to the forum (limited to 13 participants) vary from some who agree on the importance of continental evaporation to sustain regional rainfall, to others who disagree and state that continental evaporation is not important to rainfall ([1]). There is no consensus among international scientists with respect to moisture recycling, and it should be recognized as a theoretical and practical gap in the analysis of river basin scale water resources management. Enthekabi et al (1999) recognized that hydrological research at the interface between the atmosphere and land surface is undergoing a dramatic

---

[1] http://www.waterandclimate.org/MoistureWeb/Moisture_Recycling.htm

change in focus, driven by new societal priorities, emerging technologies and better understanding of the earth system.

2.6 Discussion and Conclusions

Although concrete findings showed that rainfall increases with increased land evaporation, still more research is needed to understand the cause and effect of the land surface-climate feedbacks, and to give a quantitative description of the temporal and spatial effects of land use changes on the continental rainfall. The water balance approach used in many studies to define regional moisture recycling is too simple to define accurately the physics of the land surface–climate interaction. Significant differences in the coupling strength between Soil Vegetation Atmosphere Transfer SVAT schemes and the atmospheric circulation were reported. As summarized by Brude and Zangvil (2001): *"there is plenty of room for improvement in the problem of constructing an adequate recycling model"*.

Over the land surface, the hydrological processes are also complex. Land use change, rainfall, evaporation, runoff and recharge of groundwater are mutually interrelated. Vegetation affects the hydrological cycle in several ways. It directly affects interception and transpiration as function of the leaf area, and indirectly it affects the amount of infiltration to the groundwater storage and the runoff. More methods to describe the regional scale evaporation and precipitation rates utilizing remote sensing techniques need to be explored further (see example Sudd Fig. 2.2).

Furthermore, there is an important need for research into policy implications of the link between land and water use, climate and water resources availability. An important question is what are the good land use/water resources management options that yield sustainable development and yet a positive climate feedback.

What is known and confidently accepted within the climate community about the land surface climate feedbacks processes is still not recognized and not consented within the water management community. More efforts are required to bridge the gap between the two communities during the current glooming water crisis.

3. Description of the Study area

3.1 Basic hydroclimatology of the Nile

The Nile basin covers an area of over 3 million km², and a length of about 6,700 km, longest in the world. The basin extends from 4°S to 32°N, stretching over different geographical, climatological and topographical regions (Fig. 3.1). Besides the two plateaus in Ethiopia and around the equatorial lakes (Victoria, Albert, Kayoga, Edward), the Nile Basin can be considered as a large flat plain, in particular the White Nile sub-basin.

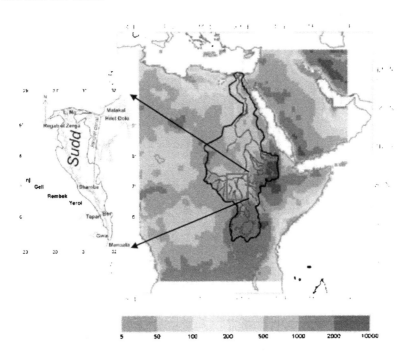

Fig. 3.1: Location and topography of the Nile Basin and Sudd wetland (m+MSL)

3.1.1.   Climate

The climate characteristics and vegetation cover in the basin are closely correlated with the amount of precipitation (Fig. 3.2). Precipitation is to a large extent governed by the movement of the Inter-Tropical Convergence Zone (ITCZ) and the land topography. The main climate zones to be distinguished from North to South are: The Mediterranean climate, a narrow strip around the Nile Delta, followed by the very dry Sahara desert climate down to around 16°N, then a narrow strip of the semi desert climate, followed by the wide Savannah climate (poor and tropical Savannah) down to the southern border of Sudan.

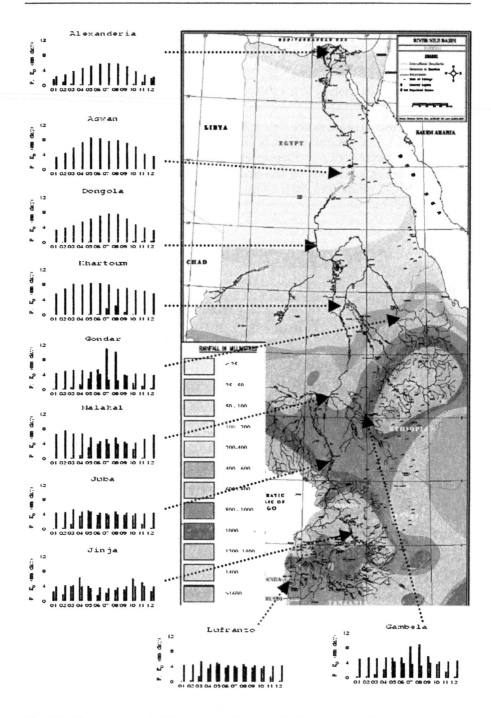

Fig. 3.2: Mean annual rainfall in mm/yr (Source: Nile Basin Atlas, TECCONILE). Mean monthly precipitation P (— Blue color), and potential evaporation E_0 (— Red color) in mm/day at key stations (source: *Smith*, 1993).

On the extreme south and southwestern boundary of the basin (around lake Victoria) tropical and rainforest climates are found. In general, precipitation increases southward, and with altitude (note the curvature of the rain isoheights parallel to the Ethiopian Plateau). Precipitation is virtually zero in the Sahara desert, and increases southward to about 1200-1600 mm/yr on the Ethiopian and Equatorial lakes Plateaus. Two oceanic sources supply the atmospheric moisture over the Nile basin: the Atlantic and the Indian Oceans.

The seasonal pattern of rainfall in the basin follows the movement of the ITCZ. The ITCZ is formed where the dry northeast winds meet the wet southwest winds. As these winds converge, moist air is forced upward, causing water vapor to condense. El-Tom (1975, p. 21) claimed that the highest precipitation falls in a region 300 to 600 km south of the surface position of the ITCZ in association with an upper tropospheric tropical easterly jet stream. The ITCZ moves seasonally, drawn towards the area of most intense solar heating or warmest surface temperatures. Normally by late Aug/early Sep it reaches its most northerly position up to 20°N. Moist air from both the equatorial Atlantic and the Indian Ocean flows inland and encounters topographic barriers over the Ethiopian Plateau that lead to intense precipitation, responsible for the strongly seasonal discharge pattern of the Blue Nile. The retreat of the rainy season in the central part of the basin from Oct onwards is characterized by a southward shift of the ITCZ (following the migration of the overhead sun), and the disappearance of the tropical easterly jet in the upper troposphere. The Inter-annual variability of the Nile precipitation is determined by several factors, of which the El Nino-Southern Oscillations (ENSO) and the sea surface temperature over both the Indian and Atlantic Oceans are claimed to be the most dominant (Farmer, 1988; Nicholson, 1996). Camberlin (1997) suggested that monsoon activity over India is a major trigger for the Jul to Sep rainfall variability in the East African highlands.

The monthly distribution of precipitation over the basin shows a single long wet season over the Ethiopian plateau, and two rainy seasons over the Equatorial Lakes Plateau as given in Fig. 3.2 for some of the key stations in the basin. Potential evaporation data E_0 (in this case equivalent to the reference crop evaporation) are also plotted. E_0 is the evaporation from a hypothetical grass crop 12 cm high with no moisture constraints, surface resistance of 70 s/m and an albedo of 0.23. The E_0 shows trends opposite to the precipitation, i.e., increases in northward direction. The climatology of the dry and hot atmosphere near Lake Aswan has a reference crop evaporation being twice the value for Upper Nile stations near Lake Victoria.

3.1.2. Hydrology and water resources

The Nile starts from lake Victoria (in fact from farther south at the Kagera River feeding the lake) and travels north, receiving water from numerous streams and lakes on both sides (Fig. 3.1, 3.2 and 3.3). In the Sudd, where it takes the name of Bahr el Jebel, the river spills its banks, creating huge swamps where more than half of the river inflow is evaporated. At Lake No, east of Malakal it is joined by the Bahr el Ghazal River draining the southwestern plains bordering the Congo Basin. The Bahr el Ghazal is a huge basin subject to high rainfall over the upper catchments, but with negligible contribution to the Nile flows. Almost all its gauged inflow (12 Gm^3) is evaporated in the central Bahr el Ghazal swamps. The Sobat tributary originating from the Ethiopian Plateau and partly from the plains east of the main river joins Bahr el Jebel at Malakal. Downstream this confluence (where it is called the White Nile), it travels downstream a mild slope up to the confluence with the Blue Nile at Khartoum. The Blue Nile originates from Lake Tana located on the Ethiopian Plateau at 1800 m above MSL, and in a region of high summer rainfall (1500 mm/yr). The only main tributary of the Nile before it ends up at the Mediterranean Sea is the Atbara River, also originating from the Ethiopian Plateau. The flows originated from the Ethiopian Plateau are quite seasonal and with a more rapid response compared to the flow of the White Nile coming from the Equatorial lakes. Further details on the Nile hydrology can be found in Shahin (1985), Sutcliffe and Parks (1999) among others.

The river catchments of the Nile tributaries were delineated based on the Digital Elevation Model (DEM) and the drainage maps of the riparian countries. The catchments areas and average annual flows are given in Table 3.1 and Fig. 3.3. The relative contribution to the mean annual Nile water at Aswan of 84.1 Gm^3 is approximately 4/7 from the Blue Nile, 2/7 from the White Nile (of which 1/7 from the Sobat), and 1/7 from the Atbara River. So the Ethiopian catchments (Sobat, Blue Nile and Atbara River) contribute to about 6/7 of the Nile water resources at Aswan.

Table 3.1: Catchment areas and mean annual flows of the sub-basins.

No.	Catchment	Outlet location	Area Gm^2	No. of model grid points	Annual Flow Gm^3/yr*
1.	Nile	Mediterranean	3310	1378	
2.	Nile	Aswan	3060	1274	84.1
3.	Atbara	Atbara	180	75	11.1
4.	Blue Nile	Khartoum	330	138	48.3
5.	White Nile	Khartoum	1730	722	26.0
6.	White Nile	Malakal	1480	615	29.6
7.	Sudd wetland	Malakal	35	14	16.1
8.	Bahr el Ghazal	Lake No	585	244	0.31
9.	Sobat	Malakal	250	104	13.5
10.	White Nile	Juba	490	205	33.3

*Mean river natural flows for the period ~1910 to 1995 (Source: *Sutcliffe and Parks*, 1999)

Ten countries share the Nile River: Burundi, Congo, Egypt, Eritrea, Ethiopia, Kenya, Rwanda, Sudan, Tanzania and Uganda. The percentage area of the Nile catchment within each country is: 0.4, 0.7, 10.5, 0.8, 11.7, 1.5, 0.6, 63.6, 2.7, and 7.4%, respectively. The Nile water is vital to the dry countries downstream (Egypt and Sudan), where historically intensive irrigation development exists, and still continues, imposing increasing demands on the Nile water. The upstream countries rely less on the Nile waters, although new water resources projects commenced in some of the upstream countries. There are bilateral and multilateral agreements between the riparian countries defining the share of the Nile water (e.g., 1891, 1902, 1906, 1925, 1929). The latest is the 1959 Nile water agreement between Egypt and Sudan (Okidi, 1990). Due to the seasonal nature of the Nile flow, several dams were built to control the Nile water for irrigation and hydropower generation. In Sudan these are: Roseires (3 Gm^3), Sennar (0.9 Gm^3) on the Blue Nile, Girba (1.1 Gm^3), on the Atbara River and the Jebel Aulia dam (3.3 Gm^3) on the White Nile. In Egypt, the High Aswan dam is by far the largest in the basin (167 Gm^3). In the upstream countries, only exist the Owen dam, which was built at the exit of Lake Victoria to generate hydropower and has no storage control. In Ethiopia there are studies for new dams across the Blue Nile, other(s) are under construction on the Atbara River basin. This huge regulation of the flows at Aswan dam constitutes a significant intervention in the natural hydrological cycle, so that it is more appropriate to consider the outflow from the basin at Dongola station (inflow to Aswan dam) rather than at the Mediterranean Sea.

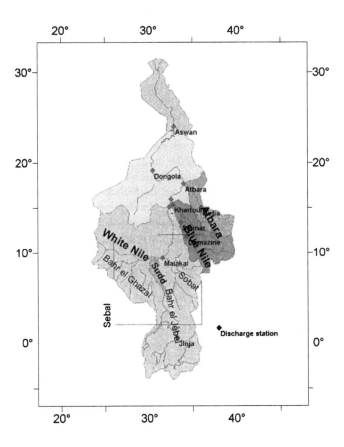

Fig. 3.3: Sub-catchments of the Nile and the discharge gauging locations.

3.2 The Sudd wetland

The Sudd wetland is one of the biggest swamps of Africa and belongs to the most extensive wetlands of the world. The Sudd is located between 6° to 9°N and 29° to 32°E, neighboring the smaller wetlands of the Bahr el Ghazal and the Machar marshes. The exact boundaries of the swamp are unspecified. The wetland of the Sudd is composed of interconnected (sometimes parallel) river channels, associated with huge flood plains. The permanent swamps, usually close to the main river courses are permanently wet. However, substantial parts of the Sudd are seasonal swamps created by flooding of the Nile or when ponds are filled seasonally with rainwater. Depending on the definition, the surface area is approximately 30,000 to 40,000 km^2. During cloud free periods remote sensing techniques are used to describe the boundaries of the Sudd system, which is a function of the Nile discharge (Travaglia et al., 1995; Mohamed et al., 2004). The area of the permanent swamps has tripled after the immense flooding of the early 1960's, which doubled the average annual inflow into the wetland as compared to the condition before 1963 (Sutcliffe and Parks, 1999). Rain falls in a single season, lasting from Apr to Nov and varying in the Sudd area from about 900 mm/yr in the south to 800 mm/yr in the north. Temperatures average to 30-33°C during the hot season, dropping to an average of 18°C in the winter season. The soils are generally clayish and poor in nutrients. The slope of the Sudd terrain is only 0.01 % and much of it is even flatter. The average annual inflow and outflow for the period 1961-83 are 49 and 21 Gm^3/yr respectively. The wetland is formed because the carrying capacity of the river channel is smaller than the incoming flow (Howell et al., 1988). Thus, the river spills over its banks, creating spill channels at different locations on both sides.

The swamps of the Sudd support a rich ecosystem. Description of the vegetation cover over the Sudd is given as part of the biophysical properties in section 5.3. The swamps and floodplains of the Sudd support rich biota, including over four hundred bird species and one hundred mammal species. Migratory birds make their stopover and wetland birds inhabit the extensive floodplains of the Sudd, while large populations of mammals follow the changing water levels and vegetation (World Wild Fund, 2001). The Sudd wetland is very important to the pastoral economy of the local inhabitants (cattle grazing).

To the west of the Sudd there is the smaller wetlands of the central Bahr el Ghazal Basin, and on the east is the Machar marshes of the Sobat River (Fig. 3.3 and 4.1). There are plans to drain part of the Sudd swamps to transfer White Nile water directly downstream. Jonglei canal phase I is one of these projects (360 km long, 2/3 completed), (Fig. 3.1). The canal aims at providing additional water of about 5% of the Nile flow at Aswan, and it may dry 30% of the Sudd swamps (JIT, 1954; Howell et al., 1988). The work on the canal stopped in 1983 due to the onset of the civil war in southern Sudan.

4. Spatial variability of evaporation and moisture storage in the swamps of the upper Nile studied by remote sensing techniques[1]

4.1 Introduction

The estimation of spatial variation of evaporation E in a catchment is fundamental to many applications in water resources and climate modeling. Evaporation, being the sum of interception, soil evaporation, open water evaporation and transpiration, is a key variable in water balance determinations, but also to estimate the moisture, heat and CO_2 interactions between land and atmosphere (see e.g., Sellers et al., 1996). Several climatic studies have indicated that atmospheric circulation and rainfall are significantly affected by the large-scale variation of soil moisture and evaporation (e.g., Savenije, 1996b; Entekhabi et al., 1999). Three methods can be used to estimate the evaporation at a regional scale: by up-scaling point measurements, by remote sensing techniques (e.g., modeling the energy balance at the land surface) and by hydrological modeling (e.g., Savenije, 1997). Each of the three methods has its limitations, and an optimal procedure probably would be a combination of the three approaches. To maximally profit from remote sensing and hydrological modeling, data assimilating is gaining terrain in hydrological studies (Walker et al., 2001; Jhorar et al., 2002; Schuurmans et al., 2003) as well as in climate studies (e.g., Dolman et al., 2001). Extensive reviews of remote sensing flux determination methods have been presented by Choudhury (1989), Moran and Jackson (1991) and Kustas and Norman (1996).

Unlike hydrological models, remote sensing techniques compute evaporation directly from the energy balance equation without the need to consider other complex hydrological processes. As a result, the error in the quantification of other hydrological processes is not propagated into evaporation E, and this is a strong advantage. On the other hand, a major limitation of remote sensing data is that the temporal distribution of satellite-based estimates is poor, and that interpolation techniques are necessary to define evaporation between satellite overpasses. Due to extremely scarce ground hydro-meteorological data in the vast study area of the Sudd marshes, utilization of satellite imagery is an attractive study approach. Advanced remote sensing techniques have a high potential to estimate hydrological processes, and are probably far better than the scanty field data available for this large tropical wetland.

---

[1] Based on: Mohamed, Y.A., W.G.M. Bastiaanssen, H.H.G. Savenije, 2004. Spatial variability of evaporation and moisture storage in the swamps of the upper Nile studied by remote sensing techniques, J. of Hydrology, 289, 145-164.

The vast wetlands in Southern Sudan are characterized by huge evaporation from the Sudd, Bahr el Ghazal and the Sobat sub-basins. The evaporation from the Sudd alone (Bahr el Jebel swamps) is estimated to be more than 50 % of the Nile inflow into the Sudd near Juba (see section 3.2 for description of the Sudd wetland). The whole river inflow of the Bahr el Ghazal Basin (12 Gm^3/yr) is evaporated before reaching the Nile. About 4 Gm^3/yr is claimed to be evaporated from the Machar marshes (PJTC, 1961). Therefore, the Sudd became subject of research by planners and engineers to save water and carry more water to the rapidly expanding population living in the downstream areas in North Sudan and Egypt. Their aim is to save water by reducing the evaporation (losses) from the wetlands. Numerous studies and projects were proposed to reduce these losses, of which the (uncompleted) Jonglei canal is the most famous (JIT, 1954; PJTC, 1961).

Most of the past studies to estimate evaporation from the Sudd wetlands rely on the computation of evaporation using meteorological ground station data under the basic assumption that the area is wet throughout the year and moisture is not limiting evaporation rates (JIT, 1954; Penman, 1963; Sutcliffe and Parks, 1999). Although some experiments were made to estimate evaporation from papyrus grown in water tanks (e.g., Butcher, 1938), their results were rejected in the subsequent hydrological studies as being too low (1533 mm/yr). Instead, more recent studies used open water evaporation of 2150 mm/yr estimated with the Penman formula (Siu-On et al., 1980; Sutcliffe and Parks, 1999). Penman (1963) assumed that transpiration from papyrus is similar to the evaporation from open lagoons. However, the computed evaporation by his well known formula represents the potential open water evaporation, and not the actual evaporation E_a from the land surface with a certain heterogeneity in vegetation types and development. The major difficulty of this estimation procedure is that evaporation from point measurements is applied to the total Sudd area. One of the complexities in determining the regional scale evaporation is the variability of the boundaries of the Sudd swamps and the soil moisture behavior throughout the year.

Some of the earlier studies intended to define the extent of the Sudd wetlands from satellite imagery. Travaglia et al. (1995) applied a methodology based on the thermal inertia difference between dry and wetlands from the NOAA-AVHRR images. NOAA-AVHRR thermal data were complemented with maps of the Normalized Difference Vegetation Index (I_{NDV}). They defined the seasonal and inter-annual variation of the Sudd area between 28.0 to 48.0 Gm^2 during 1991 to 1993. Mason et al. (1992) used the thermal channel of Meteosat to estimate the Sudd area during 1985 to 1990 between 8.0 Gm^2 at its minimum and 40.0 Gm^2 at its maximum. These methods showed a successful application of NOAA-AVHRR and Meteosat thermal data to define the area of wetlands, but such methods fail in the rainy season (cloudy conditions) and do not determine the evaporation.

The approach adopted in this study is based on the derivation of the actual evaporation using the SEBAL techniques (Bastiaanssen et al., 1998a). All components of evaporation (evaporation from open water, from vegetative surface and from bare soil) are calculated based on the surface reflectance and emittance in different parts of the spectrum. This forms basic data that can be used in hydrological and meteorological studies. SEBAL was selected because it does not require any input data, except routine weather station data.

Moisture evaporated into the atmosphere may again precipitate at a different location, and perhaps in the same basin (Savenije, 1995, 1996a; Trenberth, 1999). If this holds true, the need for reclamation efforts loose ground. Estimation of evaporation by remote sensing over the Sudd is a first step in this research to model moisture recycling in the Nile basin. The derived evaporation results from remote sensing will be used subsequently to verify moisture exchanges between land surface and atmosphere in a regional climate model (given in chapter 6).

4.2 Material and methods

4.2.1. NOAA-AVHRR images

NOAA-AVHRR images have been selected to cover 3 sub-basins: the Sudd, Bahr el Ghazal and the Sobat basins. These basins stretch from the outfalls on the White Nile to the upstream side near the discharge measuring stations. The names of the most important places are provided in Fig. 4.1.

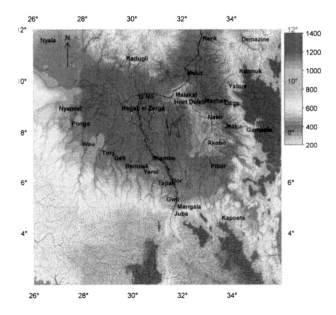

Fig. 4.1: DEM (in m+MSL), drainage map and location of hydrometeorological gauging stations

Fig. 4.1 shows the natural drainage network within the image boundary, superimposed over the digital elevation model. The largest wetlands of the Sudd are located at the center of the image, extending from Mangala in the South up to Malakal in the North. The Bahr el Ghazal basin on the west extends from the Nile-

Congo divide and joins the Sudd at several places. The Sobat Basin on the eastern side starts from the Ethiopian Plateau and joins the Nile at Malakal, it also spreads northward into the Machar marshes. The upper part of the Ghazal and the Sobat catchments lie on relatively high grounds 1000 to 2000 m above mean sea level, while the lower part, and the flat Sudd swamps lie at 300 m above MSL.

The main climatic features of the image area during the year 2000 at 4 meteorological stations are presented in Fig. 4.2a to 4.2d. The air temperature reaches its maximum in Mar/Apr and gradually declines in Jul, Aug and Sep. The annual average temperature is with approximately 28°C very high. The relative humidity has a distinct annual variation, from about 20% in the dry season to 80% in the rainy season. The reference evaporation E_0 is computed according to the FAO Penman-Monteith method (Allen et al., 1998). The monthly variations show that $E_0 \sim 10$ mm/day in the dry season and reduces to $E_0 \sim 4$ mm/day during the wet season. The accumulated values of E_0 are 2400 mm/yr for the Juba station and 2900 mm/yr for the Neyala station respectively. The actual evaporation E_a is expected to be substantially lower as the basin does not exist of a reference crop (12 cm clipped grass) with ideal moisture regimes throughout the whole year.

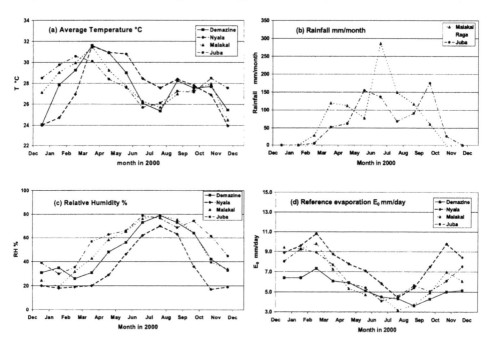

Fig. 4.2: Climatic features at Demazine, Neyala, Malakal and Juba during year 2000

4.2.2.   Major SEBAL model principles

The Surface Energy Balance Algorithm for Land SEBAL is an energy partitioning algorithm over the land surface, which has been developed to estimate actual evaporation from satellite images (Bastiaanssen et al., 1998a; 1998b). The scheme has found applications in different basins of the world, e.g., Snake River basin in Idaho, USA (Allen et al., 2002), the lake Naivasha drainage basin in Kenya (Farah, 2001), all river basins in Sri Lanka (Bastiaanssen and Chandrapala, 2003) and the irrigated Indus Basin in Pakistan (Bastiaanssen et al., 2002). In this section, a brief review of the algorithm is presented along with its assumptions, and the range of the attainable accuracies. For detailed derivation of SEBAL algorithm refer to the literature quoted above. One distinct feature of the SEBAL algorithm is that the actual evaporation from various surface types, including different leaf coverages, soil types or groundwater table conditions can be derived. SEBAL estimates the spatial variation of the hydrometeorological parameters using satellite spectral measurements and (limited) ground-based meteorological data (Farah and Bastiaanssen, 2001). These parameters of the Soil-Vegetation-Atmosphere system are used to assess the surface energy balance terms. The latent heat flux λE is computed as the residue of the energy balance equation:

$$\lambda E = R_n - G_0 - H \tag{4.1}$$

where R_n is the net radiation over the surface, G_0 is the soil heat flux, H is the sensible heat flux and λE is the latent heat flux. The SEBAL procedure consists of 25 steps, which can be re-grouped into 5 main steps appropriate for automatic processing, see Fig. 4.3. The steps are:

1. Pre-processing of the satellite image: Radiometric correction, geometric correction and removal of cloud pixels. Spectral radiances at the top of the atmosphere signaled by the satellite are converted to brightness temperatures using Planck's radiation equation and to surface reflectances. The planetary albedo derived from the visible channels 1 and 2, were radiometrically corrected to estimate the land surface albedo r_0 as in (e.g., Zhong and Li, 1988).

2. Computation of the SVAT parameters: The Normalized Difference Vegetation Index I_{NDV} normalizes the difference between red and near-infrared reflectance (Tucker, 1979). The thermal infrared emissivity ε_0 estimated based on I_{NDV} (van de Griend and Owe, 1993). The surface roughness z_{0m} estimated also based on I_{NDV} (e.g., Moran and Jackson, 1991). The temperature T_{0l} is the land surface temperature over the horizontal plain of the image which is T_0 corrected for elevation height (6.5 °C/km).

3. Computation of net radiation R_n and soil heat flux G_0. The net short wave radiation is scaled by the land surface albedo, while the net thermal infrared radiation emitted by the earth surface R_{nl} is scaled according to land surface temperatures T_0. The soil heat flux G_0 is estimated as function of R_n, T_0, r_0 and I_{NDV} over the pixels.

4. Computation of sensible heat H by iteration procedure to describe buoyancy effects on the aerodynamic resistance of the land surface r_a. First, the

temperature difference between land surface and the air dT is scaled over the image by a linear relation with T_{01}. dT is bounded by zero on a wet pixel and dT_{max} on a dry pixel selected manually over the image. The sensible heat H apart from dT is also function of r_a. However, r_a is also function of H. This is solved by iteration utilizing Monin-Obukhov similarity theory to correct for the buoyancy effects. The wind speed at the blending height u_{100}, derived from the ground measurements is used to estimate the friction velocity.

5.   Computation of instantaneous latent heat flux λE and instantaneous evaporative fraction Λ based on Eq. (4.1) and Eq. (4.2) respectively.

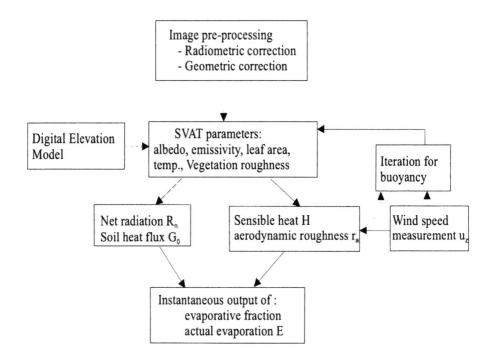

Fig. 4.3: SEBAL flow chart for instantaneous surface energy balance

The evaporative fraction, is the key parameter in SEBAL to express energy partitioning:

$$\Lambda = \frac{\lambda E}{\lambda E + H} = \frac{\lambda E}{R_n - G_0} = \frac{1}{1+B} \tag{4.2}$$

where B is the Bowen ratio ($H/\lambda E$). The advantage of using the evaporative fraction over the Bowen ratio is that the former shows less variation during the daytime than the Bowen ratio. The energy partitioning calculated with the evaporative fraction Λ

is related primarily to the available moisture content (Boni et al., 2001). Therefore, soil moisture can be empirically derived from the evaporative fraction as: (Ahmed and Bastiaanssen, 2003; Scott et al., 2003),

$$\frac{\theta}{\theta_{sat}} = e^{(\Lambda - 1)/0.421}$$

(4.3)

Where soil moisture θ is expressed as a percentage of the moisture value at full saturation θ_{sat}. The value ranges, therefore, between 0 and 1.

The main assumption to obtain daily evaporation from the instantaneous SEBAL results is that the instantaneous evaporative fraction is equal to its daily value integrated over a period of 24 hours. This was proven to hold true for the midday conditions when λE is high (e.g., Brutsaert and Sugita, 1992; Farah, 2001). The daily soil heat flux is assumed negligible as it balances out during day and night. The daily net radiation is obtained from routine meteorological data at the ground stations and then interpolated to the pixels of the image using the distributed albedo values. The daily evaporation is calculated as the instantaneous evaporative fraction times the daily net radiation. This is based on the fact that Λ is relatively constant during the day, as mentioned above:

$$E_a\big|_{day} = \Lambda * R_n\big|_{day}$$

(4.4)

The estimation of evaporation during days with no satellite image is calculated by assuming that the daily ratio of actual evaporation to the reference evaporation is valid, also on a monthly basis (Morse et al., 2000; Allen et al., 2001):

$$\frac{E_a}{E_0}\bigg|_{daily} = \frac{E_a}{E_0}\bigg|_{monthly}$$

(4.5)

Daily and monthly reference evaporation E_0 were computed by the FAO Penman-Monteith formula (Allen et al., 1998) based on routine weather data (temperature, sunshine hours, wind speed, relative humidity) at point stations. The climate data are first interpolated to the pixels of the raster maps based on spatial averages from point measurements, then E_0 is calculated assuming the land surface consists of grass. Depending on the availability of suitable images, it was attempted to prepare E_a maps for the 5th, 15th and 25th day of the month. The monthly evaporation E_a is calculated through Eq. (4.5), where daily E_a stems from SEBAL and daily and monthly E_0 from the FAO method. The presence of clouds forms part of the weather data and thus is incorporated into both $E_{0\ daily}$ and $E_{0\ monthly}$. The SEBAL procedure involves the following main requirements and assumptions, viz.:

1. The presence of a dry pixel (zero evaporation) and a wet pixel (zero sensible heat) in the same image for scaling of the sensible heat flux.

2. The wind speed at the blending height (~ 100 m) is assumed constant over the whole area of interest.
3. The temperature difference dT is a linear function of the surface temperature T_{01}.

The accuracy of the SEBAL-based evaporative fraction and daily evaporation estimates, $E_{a\ daily}$, has been evaluated in earlier studies against field measurements collected from different hydro-meteorological projects. Data from large-scale field experiments such as EFEDA-Spain, HAPEX-Niger and HEIFE-China have been explored for this. An overview of the validation results of SEBAL is given in (Bastiaanssen et al., 1998b). The error of the evaporative fraction from SEBAL single day events and for scales in the order of 1 km² is 17% or lower (at a probability of 90%) see Fig. 4.4. The error reduces when E_a is aggregated over longer time steps, because the errors cancel out, as experimentally demonstrated in the Savannah of Kenya (Farah, 2001), the humid tropical mixed vegetation of Sri Lanka (Hemakumara et al., 2003) and the irrigated Snake River basin in Idaho, USA (Allen et al., 2002).

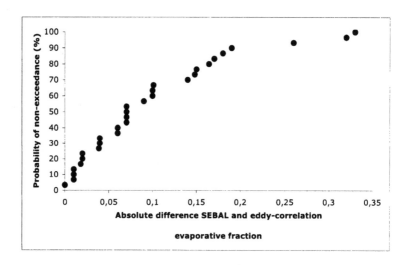

Fig. 4.4: Accuracy of SEBAL against eddy-correlation for single day events

4.3 SEBAL results

Eq. (4.1) has been applied to compute the energy balance components and hence to derive the latent heat flux on a pixel-by-pixel basis over an area of 1000 km x 1000 km, covering the Sudd and the neighboring swamps of the Ghazal and the Sobat basins. SEBAL has been applied for more than 115 images acquired during individual days of the years: 1995, 1999 and 2000, see Appendix A for the date of acquisition. The three years have different climatic conditions and river flow regimes (see Table 5.2). The spatial analysis discussed in this chapter is applied to year 2000, which has been considered to resemble average year conditions (see

Table 4.2), while the temporal analysis, which is presented in Chapter 5 covers the entire record of the 3 years.

The daily-integrated E_a values have been obtained through the evaporative fraction as defined in Eq. (4.2) and applied in Eq. (4.4). The monthly evaporation maps have been calculated from the daily evaporation maps using Eq. (4.5). Fig. 4.5 shows the results of particular places such as Neyala (dry area), Malakal (middle of the image), Sudd (center of wetland), and Juba (southern part of the image). As can be seen from these graphs, the pixels in the middle of the Sudd show the highest evaporation rates, which is in agreement with the abundant water availability present at the floor of this lowland region. Relatively lower evaporations during the rainy season were found (high humidity and more clouds). At Neyala, the evaporation is lowest, in particular during the hot dry months of Mar to Jun when the area is dry.

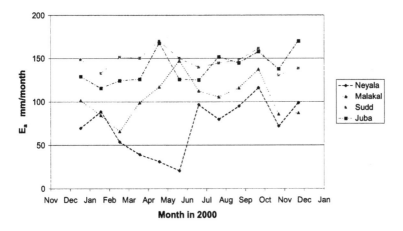

Fig. 4.5: Monthly evaporation at selected locations over the image in year 2000

It can be concluded that, during the dry season (Nov/Dec to Mar/Apr), evaporation is limited to the wetlands only, which are fed continuously from the Nile, and some little areas in the Ghazal sub-basin. During the peak of the rainy season (Aug to Oct), evaporation is high throughout the entire Sudd plain due to soil moisture saturation. Fig. 4.6 gives the annual evaporation map. It shows evaporation from wetlands (e.g., Sudd) to be higher than for other areas. Note that the white color indicates missing data (cloudy pixels). Due to uneven rainfall distribution, evaporation is lower in the northern part of the image (600 to 700 mm/yr), as well as on the southeastern corner. For the Congo catchment on the southwestern corner, the seasonal rainfall distribution is somewhat different from the Nile basin part covered by the image. I.e., the rainy season continues throughout the year, and only decreases during Dec to Feb. This is reflected in the seasonal variation of the monthly evaporation (data not shown here).

The SEBAL results can be compared to the microwave data on soil moisture indices as derived from the European Remote Sensing (ERS) satellite scatterometer data

prepared by the Institute of Photogrammetry and Remote Sensing, Vienna
University of Technology (http://www.ipf.tuwien.ac.at/radar/ers-scat/home.htm).
Although the spatial resolution given in their monthly maps is rather high (50 km),
one can see that areas of a high soil moisture index correspond to areas of relatively
high evaporation on the SEBAL maps, and vice versa.

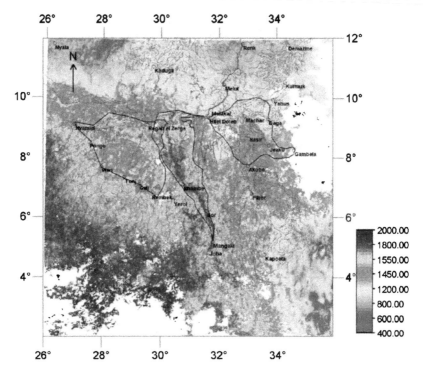

Fig. 4.6: Annual evaporation map for year 2000 (mm/yr). White colour respresents
cloudy pixels.

The seasonal variation of evaporation can also be evaluated through the variation of
the relative evaporation ratio E_a/E_0, which in the irrigation literature is known as the
crop coefficient K_c. It should be noted that E_a comes from SEBAL and E_0 from the
FAO Penman-Monteith formula (Allan et al., 1998). Fig. 4.7 shows the E_a/E_0
seasonal variation for 4 stations: Neyala, Malakal, Sudd and Juba. E_a/E_0 is around
1.0 for most of the stations during the rainy season, and it drops significantly to
about 0.5 and lower during the dry months. The Sudd pixel in the center of the
wetlands gives the highest E_a/E_0 values, while Neyala located on the driest area of
the image, shows the lowest E_a/E_0 values. In the period Mar to Apr, Neyala has a
relative evaporation as low as $E_a/E_0 < 0.20$. This clearly demonstrates that the
seasonal distribution of evaporation is related to the moisture availability and that
the surface types – except the permanently wet marshes – undergo a significant dry-
down. A maximum value of $E_a/E_0 = 1.25$ occurs in the area near the Sudd, which
shows that papyrus and other types of aerodynamically rough vegetation have an
evaporation rate more than standard clipped grass considered in the definition of E_0.

Tall crops such as sugarcane have according to generic tables of K_c a value of K_c=1.25 during the middle of the growing season (Allen et al., 1998). The latter is in good agreement with tall wetland surfaces. Chapter 5 gives a detailed assessment of the seasonal characteristics of evaporation and biophysical properties within the Sudd wetland.

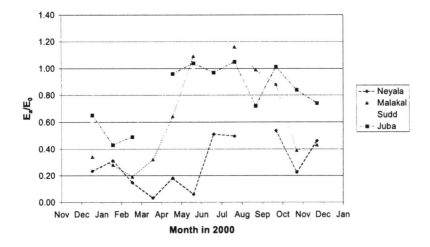

Fig. 4.7: Seasonal variation of the ratio E_a/E_0 at selected locations for year 2000

The seasonal variation of the E_a/E_0 term has important implications on the calculation of the total evaporation from the Sudd wetlands. As opposed to what is assumed in earlier studies on the Sudd with respect to permanently wet surfaces, SEBAL computes an annual evaporation of only 1636 mm/yr (computed as annual evaporation volume divided by the Sudd wet area), which is substantially less than the 2150 mm/yr for open water evaporation which was assumed to prevail throughout the wet area. In fact, Butcher (1938) conducted an experiment to determine the total annual evaporation from a water tank filled with papyrus and he found it to be 1533 mm/yr. In almost all the subsequent studies, this lower estimate has been rejected. Our conclusion is, however, that the values of Butcher are more reasonable than schematizing the Sudd to exist of permanently saturated surfaces only.

4.4 Monthly water balances

A monthly water balance has been computed for the 3 sub-basins to verify the plausibility of the new evaporation results: Sudd, Ghazal and Sobat. These sub-basins have different land use patterns, precipitation rates and accuracy of available hydrological data. Table 4.1 shows that the Sudd basin has ponding water at the surface throughout most of the area, and that apart from some higher located spots, the entire sub-basin is flooded in the wet season, or has very shallow groundwater table. The Sudd is the bottom floor of the White Nile, and the in- and outflow of the Sudd is through the Nile only. The single in- and outlet makes the area suitable for water balance determinations. The opposite holds true for the Bahr el Ghazal sub-basin, which receives inflow through numerous small streams bringing in surplus water from the upper catchment areas. The vegetation growth patterns follow the surplus flow and the impact on groundwater table fluctuations. The majority of the area has an unsaturated soil (see Table 4.1).

Table 4.1: Main characteristics of the sub-basins.

Sub-Basin	Area of sub-basin Gm^2	Catchment cover	Hydrological record
Sudd	38.6	Dominated by swamps	Good quality
Bahr el Ghazal	59.3	Mix of swamps and dry land (around 10-20 % swamps)	Incomplete, partially gauged
Sobat	42.9	Mostly dry land, with some seasonal swamps	Fair

It has to be recalled that, hydrometeorological data in the area are scanty. Many gauges are out of operation after the onset of the civil war in 1983. Therefore, it is unavoidable to use average data for some of the parameters, rather than taking data from 2000. Rainfall data for the year 2000 are available for stations; Neyala, Demazine, Renk, Kadugli, Malakal, Wau and Juba (see Fig 4.1). River flow records (water level and discharge) are not measured since 1983, with the exception of: Malakal, Renk and Demazine, located north of the Sudd. The comparison of the Nile annual flow at Malakal for the year 2000, and the long term mean shows negligible difference, while the 2000 rainfall over the Sudd is slightly less than the long term mean, (Table 4.2). It is anticipated beforehand that a small error in the water balance is expected by assuming the river inflow, outflow and monthly rainfall of 2000 to be similar to the average condition.

Table 4.2: Comparison of year 2000 and long-term average of 1960 – 2000.

Parameter	Year 2000	Long term average (1961-2000)
Nile Flow at Malakal	33.84 Gm^3/yr	32.04 Gm^3/yr
Rain average of Malakal, Juba and Wau	950 mm/yr	923 mm/yr

4.4.1.   The Sudd water balance

The areal size of the sub-basins is one of the key problems for assessing the water balance. The Sudd swamps results from water spillage on both sides of the Bahr el Jebel River, and it extends from near Juba up to the confluence with the Sobat River just upstream Malakal. There is an ongoing debate on the catchment boundaries, which cannot be straightforwardly surveyed because of its immense dimensions and because the area is not freely accessible. The area of the Sudd swamps shrinks and swells during the season. The boundary between the Sudd and Bahr el Ghazal swamps is highly questionable, and some hydrological state parameters should be used to help identifying the boundaries. The catchment boundary has – in this study – been delineated based on the annual evaporation map of the average year because wetlands can be discerned from its surrounding areas (see Fig. 4.6). The annual evaporation map can be considered as a suitable indicator for the annual wetland area. A threshold of annual evaporation of 1550 mm/yr is selected to represent open water bodies covered with swampy vegetation. The high annual evaporation in the area can be attributed to the presence of papyrus and water hyacinth grown on water, and dense flooded woodlands. The delineated Sudd area is 38.3 Gm^2, which is 74% larger than the most recent estimate of 22 Gm^2 (average area) by Sutcliffe and Parks (1999).

The hydrological records of inflow, outflow and rainfall – used in this study to verify the evaporation estimates from remote sensing – are based on long term averages and are withdrawn from Sutcliffe and Parks (1999). The average rainfall from 8 stations in the area was used. Inflow is taken as the river flow at Mangala. The two tributaries (Lau, Tapari), coming from the west entering the Nile north of Mangala, are not included in the balance as their flow is small (1.6 Gm^3/yr, compared to the Nile inflow of 49.6 Gm^3/yr), and most likely a high proportion of this water spreads and evaporated before reaching the Sudd. It is relevant to review the water balance of the Sudd made by Sutcliffe and Parks (1999) to understand their model concept and interpret the value of their results. Their formulation of the water balance was as follows:

$$\mathrm{d}S = [R_{in} - R_{out} + A(P-E) - r\,\mathrm{d}A]\mathrm{d}t \qquad\qquad (4.6)$$

where S is the storage, R_{in} is inflow, R_{out} is outflow, P is rainfall, E is evaporation, A is the flooded area and r is soil moisture recharge. This term accounts for the infiltration in the unsaturated zone during the expansion of the swamps. There is no capillary rise (negative recharge) foreseen when the groundwater table draws down. However, this term is negligible compared to the other terms. Note that the recharge pertains to expanding areas dA only. A linear relation between A and S is assumed to eliminate two unknowns from Eq. (4.6). The area ponded with water A and the surface storage S are schematized to be linearly related as $A=kS$, with a constant depth $1/k=$ 1m. The evaporation E was assumed to be from open water surfaces and was computed by means of the Penman formula using meteorological data from Bor. Penman (1963) assumed that transpiration from papyrus is similar to the evaporation from open lagoons. The resulting water balance for open water areas as a result of Eq. (4.6) is provided in Appendix B (Table B1). Iterative calculations have been made to close S at the start of the year with S at the end of the year. A key result is that an average swamp area of 21.1 Gm^2 is required to conserve the water balance based on an annual open water evaporation of 2150 mm/yr.

Satellite imagery of the Normalized Difference Vegetation Index I_{NDV} shows, however, that wetlands in the Sudd are covered with vegetation: papyrus, water hyacinths, etc. (I_{NDV} values ranges between 0.2 to 0.3 over the Sudd), while water bodies and reservoirs have I_{NDV} below zero (-0.2 to –0.3). Therefore, the evaporation schematization of the Sudd to exist of open water surface is basically incorrect. Secondly, evaporation from the papyrus grown in water is less than open water evaporation, because the albedo of green vegetation is (theoretically) 15%, 3 times higher than an albedo of 5% for water, and this yields to less net radiation available for E_a. The actual values of albedo derived over the image were 0.15 to 0.25 over the Sudd, while it is around 0.10 to 0.15 over the Demazine reservoir. In addition, a canopy resistance is present in plants due to the pathway of water flow through the stems and the obstruction for water vapor through stomata, which does not hold for open water surfaces. The evaporation resistance to papyrus and water hyacinth is therefore in all circumstances higher than for open water where water molecules can freely convert from liquid into gaseous phase. Hence, open water evaporation should be higher than the evaporation from vegetative wetlands. However, in the literature (e.g., Gilman, 1994) the comparison of wetland evaporation against open water evaporation shows a wide range of results, and the truth is probably very site specific, as it depends on the unique combination of the soil, climate and the bio-physical parameters including albedo, I_{NDV}, aerodynamic and canopy resistances. Detailed evaluation of wetland evaporation and biophysical properties in relation to an open water body is given in section 5.2.

The evaporation volume from all pixels within the delineated sub-basin of the Sudd has been computed on a monthly basis. Applying the water balance based on the monthly evaporation volume as:

$$\frac{dS}{dt} = (R_{in} + P) - (R_{out} + E) \tag{4.7}$$

where dS/dt is the change in storage, gives results of Table 4.3. The accumulated value for 12 months of dS/dt=-1.15 Gm^3/yr is very small and less than 2% of the annual evaporation. This finding supports two conclusions: (i) the Sudd area is within 38 Gm^2, approximately 74% larger than the 21 Gm^2 assumed in previous studies and (ii) the evaporation rates are significantly lower as assumed previously due to foliage properties and an annual soil moisture cycle.

Table 4.3: Water balance of the Sudd basin (Area = 38.6 Gm^2)

Month	P mm/month	E (SEBAL) mm/month	R_{in} Gm^3/month	R_{out} Gm^3/month	dS/dt Gm^3/month
Jan	2	141	3.89	1.95	-3.44
Feb	3	125	3.37	1.63	-2.96
Mar	22	135	3.62	1.77	-2.51
Apr	59	141	3.60	1.62	-1.19
May	101	146	4.00	1.58	0.67
Jun	116	138	3.86	1.50	1.51
Jul	159	129	4.19	1.51	3.83
Aug	160	143	4.65	1.59	3.69
Sep	136	135	4.61	1.70	2.95
Oct	93	150	4.69	1.94	0.58
Nov	17	121	4.47	1.94	-1.50
Dec	3	131	4.21	2.06	-2.79
Total	871	1636	49.16	20.80	-1.15

4.4.2.   The Bahr el Ghazal water balance

The streams of Bahr el Ghazal Basin start from the Nile Congo divide at an elevation of 700 to 1000 m, and flows in the northeastern direction. Rainfall over the upper catchment is not well gauged, but it is assumed to vary between 1200 to 1600 mm/yr, and declines to about 900 mm/yr on the plains (see Fig. 3.2). There is sufficient slope to create drainage up to the line of the gauging stations: Juba, Rembek, Wau and Nyamlel (location shown in Fig. 4.1), where numerous un-gauged streams flow on very mild slopes and cause spillage over the banks. The gauging stations roughly follow the boundary of the ironstone plateau, and thus lie at the edge of the zone of runoff generation. Out of the total inflow of 11.3 Gm^3/yr measured at the upstream stations, only a small fraction of 3% emerges at the basin exit just upstream Lake No.

The Bahr el Ghazal sub-basin has been delineated on the basis of streams and location of gauging stations (see Fig. 4.6). The boundaries are hydrologically correct, but the total area could be easily expanded if more streams are included. The gauging stations used for the delineation are Nyamlel on the Lol River, the Road Bridge across the Pongo River, Wau on River Jur, Tonj on River Tonj, the Road Bridge across the Maridi River near Rembek and Mvolo on River Naam. Although these streams have been gauged (intermittently) starting from the 1930's and 40's, measurements at high flows are very limited. It is probable that during high flows these stations underestimate flows that bypass the gauge over the inundated land. The rivers flowing to Bahr el Jebel: Lau, and Tapari are excluded from the balance. The total area within the catchment boundary amounts to 59.2 Gm^2 (see Table 4.1). The flow and rainfall data given in Table 4.4 are withdrawn from Appendix B (Table B2).

Table 4.4: Water balance of the Baher el Ghazal sub-basin (Area = 59.37 Gm^2).

Month	P mm/month	E (SEBAL) mm/month	R_{in} Gm^3/month	R_{out} Gm^3/month	dS/dt Gm^3/month	dS/dt (θ) Gm^3/month
Jan	0	124	0.05	0.02	-7.30	-3.63
Feb	4	106	0.01	0.03	-6.05	-0.40
Mar	14	104	0.00	0.04	-5.35	-3.31
Apr	49	130	0.01	0.04	-4.84	2.85
May	110	135	0.16	0.03	-1.37	2.05
Jun	143	126	0.47	0.02	1.47	-0.83
Jul	175	118	1.06	0.02	4.43	3.23
Aug	184	142	2.01	0.03	4.47	1.12
Sep	141	136	3.05	0.02	3.32	-1.17
Oct	69	139	2.92	0.02	-1.22	1.87
Nov	10	113	1.29	0.01	-4.85	-4.83
Dec	1	127	0.31	0.01	-7.19	3.04
Total	900	1499	11.33	0.31	-24.48	0.00

The annual evaporation of Bahr el Ghazal is 1499 mm/yr (see Table 4.4), which is less than found for the Sudd area and in agreement with the large tracts of unsaturated soil present during the dry season. The monthly water balance using SEBAL evaporation volumes formulated according to Eq. (4.7) shows a shortfall of dS/dt=-24.5 Gm^3/yr. This shortfall can be explained by two possibilities: the inflow should be considerably more (by a factor 3), or there is a carry-over storage. However, storage changes over a period of a year are expected to be small. To test the latter, dS/dt was computed from the changes in soil moisture storage in the upper 1 m of the unsaturated/saturated soil column. Eq. (4.3) has been used to compute the degree of soil moisture saturation (θ/θ_{sat}) for every month. Taking a soil moisture content at full saturation of 0.45 m^3/m^3, being an average value for alluvial soils, the opportunity is provided to compute the month-to-month variation of soil moisture storage for a soil layer of 1 m depth, Eq. (4.8).

$$S = \frac{\theta}{\theta_{sat}} \times 0.45 \times 1 \qquad (4.8)$$

An example of the monthly variation of dS/dt based on soil moisture computations is shown in Fig. 4.8. The result of Fig. 4.8 shows a distinct temporal variability, but confirms that the annual storage change is negligibly small (see also Fig. 4.9 for the dynamics of one specific pixel).

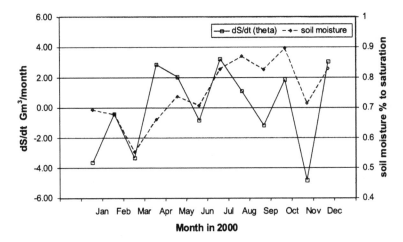

Fig. 4.8: Bahr el Ghazal sub-basin: average monthly soil moisture ($\theta/\theta sat$) and change of storage (dS/dt) in the unsaturated zone during year 2000.

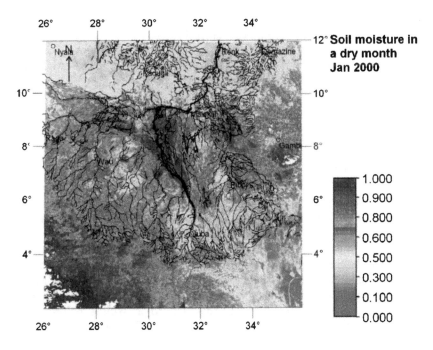

Fig. 4.9: soil moisture ($\theta/\theta sat$) map of January 2000

Since the missing inflow of -24.5 Gm^3/yr cannot be attributed to rainfall anomalies and storage releases, it has to be concluded that the error is related to the un-gauged or inadequately gauged runoff from the higher catchment to the lower Ghazal plains. A similar shortfall was computed in the balance derived by Chan and Eagleson (1980) for the Ghazal basin, who also explained un-gauged runoff as a missing element in the hydrological puzzle, plus 6 Gm^3/yr, they assumed as spillage from Bahr el Jebel to the Ghazal swamps. Spill from Bahr el Jebel to the Ghazal basin is likely to be small, since there are no visible channels and also that the balance of the Sudd alone fits as given in section 4.4.1. Chan and Eagleson (1980) in their water balance calculation of the Ghazal central swamps, estimated additional unmeasured inflow as 19.8 Gm^3/yr. Their amount of missing water (plus 6 Gm^3/yr, assumed spillage from Bahr el Jebel) is in fair agreement with the 24.6 Gm^3/yr found in this study. The interim conclusion related to the validation of the Ghazal basin is that this basin is not suitable for testing. The runoff from the upper catchment should be about 3 times higher than is actually recorded.

4.4.3.   The Sobat water balance

It has to be noted that, inflow and outflows to the Machar swamp area, used in earlier studies differ appreciably from one study to another. The Jonglei investigation team (JIT, 1954) estimated annual spill of 2.8 Gm^3/yr from Baro towards Machar Marches, plus annual flow of 1.7 Gm^3/yr from the eastern catchment, and an outflow of 0.5 Gm^3/yr. In the study of El-Hemry & Eagleson (1980), runoff to the Machar marshes from the eastern torrents and plains was estimated as 5.61 Gm^3/yr, plus 3.5 Gm^3/yr as spill from Baro, outflow from Machar estimated as 0.12 Gm^3/yr. Sutcliffe and Parks (1999) in their investigation of the flow records between 1950 to 1955, estimated flows to the Machar marshes as: 2.3 Gm^3/yr spill from Baro, 1.7 Gm^3/yr from eastern streams, and outflow of 0.12 Gm^3/yr. The Sobat contributes to about half of the White Nile flows at Malakal (13 Gm^3/yr). The average rainfall over the Machar is 933 mm/yr.

Rainfall over the basin varies appreciably, around 800 mm/yr on the plains to 1300 mm/yr near Gambela, and it is even higher on the Ethiopian mountains. The average rainfall over the delineated catchment is estimated as the average of Gambela and Machar marshes (1115 mm/yr), which is more than the 871 mm/yr for the Sudd, see Appendix B (Table B3). At large discharges, the Baro River spills over towards the Machar marshes on the North, which receives also flow from smaller streams from the east. The river system of the Sobat, and down to the Machar marshes is very complex and little is known to determine an accurate water balance. There are a few gauging stations in the Basin, but they are not operational.

The catchment boundary given in Fig. 4.6, has been delineated from the gauging stations of: Akobo, Gambela, Daga and Yabus up to the confluence near Malakal. The outflow is defined at the Sobat mouth (Hillet Dolieb). Outflow to the Nile north of the Machar marshes through Khor Adar and Khor Wol is believed to be negligible. The inflow constitutes river discharges from: Baro at Gambela, Pibor at Akobo, Akobo at Akobo, Gila, Mekwai, Jeakau, and the eastern streams of Dagga and Yabus. About 13 Gm³/yr comes from Baro at Gambela, and only 5 Gm³/yr comes from the other streams (Table 7.2 in Sutcliffe and Parks, 1999).

The results of the water balance are provided in Table 4.5. On an annual basis, dS/dt computed according to Eq. (4.7) becomes dS/dt= -3.1 Gm³/yr. The monthly variation of dS/dt based on soil moisture computations shows distinct temporal variation in the Sobat sub-basin, but confirms negligible annual change of storage. The annual dS/dt closure term amounts to 5.7 % of the annual evaporation within the confidence limits of the collected longer-term average flow data. Fig 4.6 shows no wetlands in the Sobat catchment similar to the Sudd and Ghazal. The famous Machar marshes appear to be seasonal wetlands whose evaporation drops drastically during the dry season, yielding a substantially smaller annual evaporation than the Sudd and Ghazal swamps.

Table 4.5: Water balance of the Sobat sub-basin (Area = 42.9 Gm²)

Month	P mm/month	E (SEBAL) mm/month	R_{in} Gm³/month	R_{out} Gm³/month	dS/dt Gm³/month	dS/dt (θ) Gm³/month
Jan	0	121	0.35	1.02	-5.85	0.10
Feb	2	102	0.23	0.45	-4.51	0.33
Mar	4	84	0.22	0.29	-3.53	-5.08
Apr	37	91	0.28	0.25	-2.29	0.97
May	130	115	0.62	0.43	0.84	3.79
Jun	151	113	1.57	0.86	2.33	-0.58
Jul	214	111	2.65	1.29	5.79	2.57
Aug	288	117	3.53	1.59	9.30	1.22
Sep	166	125	4.05	1.77	4.04	0.50
Oct	92	128	2.75	1.99	-0.76	0.10
Nov	31	83	1.11	1.98	-3.11	-5.02
Dec	0	98	0.60	1.76	-5.35	1.12
Total	1115	1287	17.96	13.69	-3.12	0.00

4.4.4. The total drainage system West from the White Nile

The western part of the White Nile has only a minor contribution to the flow of the River Nile. The outflow of the Bahr el Ghazal sub-basin is 0.31 Gm³/yr only. This implies that rainfall and evaporation for the natural drainage system west of the Nile must be in approximate equilibrium. This test was worked out further by plotting the outer catchment boundaries on the drainage network map. The Ghazal catchment, from the Nile-Congo divide- below Bahr al Arab and west of the Sudd was determined (236.5 Gm^2). The area above Bahr el Arab was excluded as the inflow of Bahr el Arab and Ragbat el Zaraga is negligible (0.5 Gm^3/yr). Average evaporation calculated with SEBAL over the catchment amounts to 1580 mm/yr, while average rainfall, derived from Fig. 3.2 amounts to 1200mm/yr. This shows a mismatch of 380 mm/yr. Secondly, the catchment area of both Bahr el Ghazal (below Bahr el Arab) and the Sudd west of Sobat, was determined as 350.5 Gm^2. Average evaporation and rainfall over this catchment, amounts to 1616 mm/yr, and 1150 mm/yr respectively. Inflow is 49.2 Gm^3/yr, and outflow is 20.8 Gm^3/yr. The balance of this catchment shows a mismatch of 385 mm/yr. There are two possible explanations for this mismatch: either the rainfall over the highland of the Bahr el Ghazal catchment is underestimated, or, SEBAL overestimates evaporation over the highlands, mainly composed of dense forests. Since SEBAL gives acceptable results for the other investigated basins, it is more likely that rainfall is underestimated, in particular from the forested highlands. It is very well possible that rainfall on the highlands, which is not being recorded, is substantially higher than is indicated by the isohyets of Fig. 3.2.

4.5 Conclusions

Monthly actual evaporation and soil moisture maps have been computed for an area of 1000 km x 1000 km of the Nile basin covering the wetlands of the Sudd, the Bahr Ghazal, and the Sobat. The SEBAL algorithm has been used to derive the energy balance components at the land surface from the NOAA-AVHRR images because only routine weather data need to be known. The evaporation estimates were verified against water balance data of year 2000 (resembles average condition), and close match was derived for two of the three selected sub-basins. The closure terms in the water balance were 1.8% and 5.7 % for the Sudd and the Sobat sub-basin. The balance does not close for the Bahr el Ghazal sub-basin, due to considerably small inflow, attributed to un-gauged catchment runoff.

The areal average evaporation rates for the three sub-basins are 1636 mm/yr (Sudd), 1499 mm/yr (Ghazal) and 1287 mm/yr (Sobat) which is substantially less than the 2150 mm/yr used in earlier studies (see table 4.6). These differences can be explained by the inclusion of larger areas not permanently saturated throughout the year. In the Sudd, the difference is caused by both the dominant evaporation mechanism and the areal extent of the swamp. The value used for evaporation by Butcher of papyrus (1533 mm/yr) is in good agreement with the SEBAL estimates of the Sudd. The analysis showed that the Sudd wetlands are larger than assumed before and that these areas contain moist soil with shallow ground water tables. Comparison of the monthly (actual) evaporation, E_a against reference evaporation E_0, and its seasonal variations showed that, evaporation from the three sub-basins is largely controlled by the soil moisture condition. A summary of historical evaporation assessment studies in the Sudd is given in Table 4.6.

Table 4.6: Different estimates of evaporation over the Sudd swamp.

Evaporation (mm/yr)	Average Sudd Area (Gm^2)	Source	Method
1,533	7.2	Butcher, 1938	Measurements of Papyrus grown in water tanks, areal photo, water balance
	8.3	Hurst and Black, 1931	water balance and limited surveys
2,400		Migahid, 1948	Lysimeter experiment on the Sudd, close to Bahr el Zeraf cuts.
2,150	21.1	Sutcliffe and Parks, 1999	Penman formula, water balance
1,636	38.0	This study	Remote sensing and SEBAL

5. The temporal variation of evaporation from the Sudd wetland and its biophysical and hydrological interpretation[1]

5.1 Introduction

Wetlands are distinct parts of the earth's ecosystem characterized by the absence of water constraints and rich flora and fauna. Wetlands have many beneficial functions: among others they provide: (i) flood mitigation, (ii) storage, (iii) groundwater recharge, (iv) a bird and wild life habitat, and (v) carbon sequestration, to mention a few. Globally, wetlands vary in size from the huge Russian wetlands (nearly 1 million km^2) to the small natural and man made wetlands of a few km^2. The Sudd (38,500 km^2) is one of the largest wetlands in Africa.

About half of the world wetlands are threatened by a shortage of water supply, and more surface water resources should be committed to keep these areas wet and green. An accurate determination of the water balance components (precipitation, evaporation, inflow, outflow and interaction with groundwater) is essential for determining environmental flows to keep these systems healthy. Usually evaporation from a wetland (including: open water evaporation, plant transpiration and wet/dry soil evaporation) is a major component of its water budget, though complex to determine (Linacre et al., 1970; Lott and Hunt, 2001).

World wide, there are numerous field experiments executed to measure and model wetland evaporation E_a. However, the results remain site-specific and are difficult to extrapolate into the regional context (e.g., Lafleur and Rouse, 1988; Souch et al., 1998). Most earlier estimates of E_a, assume that E_a from healthy wetlands resembles open water evaporation E_w. Penman (1963) pointed out that swamp evaporation measurements in the Sudd correspond with estimates of open water evaporation. There are also authors who assume that E_a resembles the potential evaporation E_p, i.e., evaporation from vegetative cover with no water constraint, e.g., Lott and Hunt (2001).

A wetland system is a mixed composition of marshland vegetation types, open water bodies and (un)saturated soil. Depending on the vegetation canopy structure (Leaf Area Index, and vegetation height), the wetland vegetation may intercept the incoming solar radiation, and can shelter the blowing wind. If the vegetation growth is intense, i.e., with large leaf area index and an aerodynamically rough surface, transpiration may be high. A classical wetland evaporation research question is:

[1] Based on: Mohamed, Y.A., Bastiaanssen, W.G.M., Savenije, H.H.G., van den Hurk, B.J.J.M. 2004, The temporal variation of evaporation from the Sudd wetland and its biophysical and hydrological interpretation determined by remote sensing techniques, submitted to J. of wetlands.

Does the transpiration provided by the wetland vegetation offset the deficit caused by the vegetation shading or exceeds it (Gilman, 1994)? This question can be re-phrased into; how does the surface energy balance change when open water is gradually superseded by vegetation coverage? In this chapter we review the value of the biophysical parameters that affect the E_a estimates over wetlands in general, and re-examine the value of the E_a/E_w ratio from a theoretical point of view.

Determination of E_a in the Sudd area by remote sensing and a surface energy balance model is an attractive methodology, because ground data from the area is not available, not even a pan evaporation value (Chapter. 4). This chapter discusses the biophysical properties and related surface fluxes of the Sudd wetland, determined from the SEBAL remote sensing technology (section 4.2), in relation to in- situ measurement conditions on similar habitats elsewhere in the world. The objective is to further validate the remote sensing derived parameters and interpret their seasonal characteristics.

5.2 Evaporation from Wetland

5.2.1.   Estimation of wetland evaporation

The evaporation rate from a wetland, E_a, depends on the atmospheric demand, the biophysical characteristics of the vegetation (radiation properties and resistance to evaporation) and the soil water potential in the root zone of marshland vegetation. In a healthy wetland, soil moisture is no binding constraint and the soil water potential is low. If the water table drops below the root zone in the dry season due to insufficient river flow, evaporation can become moisture controlled.

In general, wetland evaporation is estimated based on either direct measurements or through modeling. In the literature there are examples of water balance and energy balance techniques (see Eq. 4.7, and Eq. 4.1). The Bowen ratio method is used most frequently in the energy balance approach (Rinks, 1969; Burba et al., 1999; Josè et al., 2001). Eddy correlation techniques were used in some other cases (Souch et al., 1998; Campbell and Williamson, 1997; Jacobs et al., 2002). Applications of remote sensing in wetlands – also categorized as energy balance techniques – exist but are very limited. Bauer et al. (2002), for instance, computed the evaporation of the Okavango swamps in Botswana using the SEBAL methodology. They used a 1 km grid for AVHRR data and explored the interaction between surface moisture and saturated groundwater systems.

The water balance approach to estimate wetland evaporation remains an instrumental technique, in particular when the other components of the balance (precipitation, runoff, interaction with groundwater) are accurately defined. Lysimeter and water tank experiments are based on water balance principles. Lysimeter measurements of wetland evaporation are not common (e.g., Lott and Hunt, 2001). Many of the older experiments on wetland evaporation are made through water tank measurements (e.g., Butcher, 1938; van der Weert and Kamerling, 1974).

The use of formulae to estimate wetland evaporation from routine meteorological data is an interesting approach, in particular after calibration on field measurements. Some authors – in particular during the early decades – believed that wetland evaporation E_a is similar to open water evaporation E_w, and hence it can be determined using an open water evaporation formulae (e.g., Penman 1963). Examples of evaluating wetland evaporation with the Penman open water formulae are given in Souch et al. (1998) and Koerselman and Beltman (1988) among others. Morton (1983) uses the so-called CRLE (Complementary Relationship Lake Evaporation) model to estimate shallow lake evaporation. The method automatically considers the effect of different surroundings (oasis effects), but ignores the effects of varying wind speed. The original Priestley-Taylor equation with α coefficient of 1.26 was intended for partitioning net radiation into latent and sensible heats over substantial saturated land areas (Priestley and Taylor, 1972). Souch et al. (1998), and Jacobs et al. (2002) show a review of Priestly-Taylor applications in wetlands with the α coefficient ranging from 1 to 1.26, confirming that the theory of atmospheric feedback suggested by Priestley and Taylor can be applied. Note that if α is unity then evaporation is the equilibrium evaporation determined only by the available energy. Although the Penman-Monteith (P-M) equation is developed for agricultural crops, several applications to determine E_a in a wetland with the P-M equation have been established (e.g., Souch et al., 1998; Jacobs et al., 2002). The main advantage of the P-M model is that once the vegetation biophysical properties are known, detailed field measurements can be avoided. A potential difficulty of the P-M model is the prediction of heterogeneous vegetation and soil moisture conditions. However, this data demand can be met with remote sensing technologies.

5.2.2. Wetland evaporation versus open water evaporation

If a wetland is replaced by an open water body, will the water evaporation E_w be equal, smaller or larger than the wetland evaporation E_a? A review of evaporation E_a from wetlands throughout the world reveals large variation, obviously reflecting a spectrum of climate conditions and biophysical properties, as well as the intra-seasonal fluctuations of soil moisture. Normalizing E_a with open water evaporation E_w attains an opportunity to exclude the effects originating from climatic factors. Appendix C (Table C1) gives a review of the average E_a/E_w measurements for different wetlands in the world. Vegetation type and method of measurement are defined for each case. In addition, Gilman (1994) shows a wide variation of E_a/E_w between 0.85 and 2.5 of selected wetlands. Campbell and Williamson (1997) show for 4 locations that E_a/E_w varies between 0.74 to 0.90. Dolan et al. (1984) in his review of the studies made in the 1960's and 70's shows that E_a/E_w ranges from 0.1 to 4.0. Linacre et al. (1970) reviewed E_a/E_w to lay between 0.6 to 2.5. Other authors also show different results of E_a/E_w (e.g., Idso and Anderson 1988; Abtew and Obeysekera, 1995). There are two interesting reports made by Lafleur and Rouse (1988) and by Berger et al. (2001), on a transect of evaporation measurements across wetlands covering open water, vegetation on water, and vegetation on the border with relatively drier soil wetness. The climatic condition is more or less similar over the transect of the wetlands, nevertheless the results confirmed that the presence of the vegetation canopy in the wetland markedly reduced the evaporation efficiency compared to the open water site, and that the border (drier) part of the wetland has lowest E_a/E_w. Also Nieveen (1999) investigated how the evaporation changes in a landscape that contains a sequence of patches of peat bog, grass, forest and agricultural land across a distance of 6 km in The Netherlands. The lesson learned from this review is that wetlands demonstrate a substantial spatial variation and that E_a/E_w is not equal to unity.

When evaluating the results of Appendix C (Table C1), one should consider the possible measurement limitations, which may have had direct impact on the results. Dolan et al. (1984), Gilman (1994) and Souch et al. (1998), among others, attributed the possible reasons of the wide range of E_a/E_w to be related to measurement limitations, e.g., duration of the measurements and location of the site relative to the surroundings (oasis effects, land sea breeze). Van der Weert and Kamerling (1974) attributed the high ratio of E_a/E_w of 3.7 for water hyacinth, obtained by Timmer and Weldon (1967), and of 3.2 obtained by Penfound and Earle (1948) to the border effects. The very low value of E_a/E_w over the New Zealand wetland was claimed to be attributed to the severely restricted E_a by low nutrient soil conditions (Campbell and Williamson, 1997). The position of the ground water table relative to the root depth – however – has direct control on the wetland E_a as demonstrated by Jacobs et al. (2002) and Kim and Verma (1996), as expected on the basis of standard soil physical mechanisms.

In the following we will address the reported range of E_a/E_w as a result of a wide range of biophysical properties determining E_a/E_w.

5.2.3.   Bio-physical properties of wetlands

Knowledge on the wetland biophysical properties is a pre-requisite for a better understanding of the variability of wetland evaporation. There is a tantamount of literature on these properties for dry vegetation, in particular for agricultural crops (e.g., Jensen, 1980; Jarvis, 1976), while research on wetland evaporation in an international context is rather limited (e.g., Burba et al., 1999; Koch and Rawlik 1993). Plant characteristics and their impact on the canopy resistance and aerodynamic resistances r_c and r_a respectively, not to mention radiation properties of wetlands (albedo, fraction of absorbed photosynthetical active radiation and thermal infrared emissivity) are key for explaining the range of E_a values. To understand the behavior of the controlling factors on E_a let us explore the theoretical background using the P-M as given by:

$$\lambda E = \frac{\Delta(R_n - G) + c_p \rho_a \dfrac{(e_s - e_a)}{r_a}}{\Delta + \gamma(1 + \dfrac{r_s}{r_a})} \qquad (5.1)$$

where λE is the latent heat flux (λ is the latent heat of vaporization), Δ (kPa/C°) the slope of the saturated vapor pressure curve, R_n the net radiation, G_0 the ground heat flux (including heat stored in the water table), c_p (MJ/kg/C°) the specific heat at constant air pressure, ρ_a (kg/m^3) the air density, (e_s-e_a) the vapour pressure deficit (kPa), r_a (s/m) the aerodynamic roughness, γ (kPa/C°) the psychrometric constant, and

r_s (s/m) is the bulk surface resistance. The latter is a mixture of canopy resistance r_c (that dictates canopy transpiration), soil resistance (that controls soil evaporation) and the resistance for open water (usually zero when the water body is unpolluted). The bulk surface resistance represents a heterogeneous wetland ecosystem and is equal to r_c if soil and water surfaces are completely covered.

If the climatic factors Δ and (e_s-e_a) can be assumed to be constant over both a water body and a wetland, then R_n, G_0, r_s and r_a, are the remaining wetland parameters that affect λE. For a water body r_s is zero, and the liquid particles are transferred freely into water vapor without passing through stem and xylem. On the contrary, wetland vegetation has a lower r_a, which introduces a stronger coupling with the atmospheric boundary layer than a smooth water surface (McNaughton and Spriggs, 1986). Thus, vegetated wetland surfaces have a higher r_s than a water body, but a lower r_a value, and these factors have compensating effects on evaporation. The question therefore is: what is the combined r_a-r_s effect in conjunction with the variability of net available energy (R_n-G_0) on λE? Before discussing this further, some more background information is provided.

Numerous researches have been spent on simplifying the computation of soil heat flux, rather than using soil thermal properties and highly variable soil temperatures. An alternative to the classical heat conduction equation is an empirical estimation of the G_0/R_n ratio. Field measurements have indicated that in very hot deserts this ratio is approximately 0.4 and that for vegetated soil with light interception the ratio hardly exceeds 5 to 10% during daytime hours. It is customary to apply a Leaf Area Index dependent light extinction parameterization of the G_0/R_n ratio (Choudhury et al., 1987).

The aerodynamic resistance r_a depends on wind speed u_z, vegetation structure (leaf area index and vegetation height) and buoyancy effects. For neutral atmospheric boundary layer conditions (sensible heat flux over vast wet terrain is usually small), r_a can be calculated as (Monteith and Unsworth, 1990):

$$r_a = \frac{\ln\left[\dfrac{z-d}{z_{0m}}\right]\ln\left[\dfrac{z-d}{z_{0h}}\right]}{k^2 u_z} \tag{5.2}$$

where z (m) is the reference height of wind measurements, d (m) the displacement height, z_{0m} (m) the roughness height for momentum transfer, z_{0h} (m) the roughness height for heat and vapor transfer and k is von Karman's constant. For a wide range of vegetations z_{0m}, z_{0h} and d can be estimated as $0.123h$, $0.1z_{0m}$ and $^2/_3h$ respectively, where h is the plant height (Allen et al., 1998). A more comprehensive prediction of z_{0m} and d in Eq. (5.2) is made feasible by the analytical expressions of for instance Perrier (1982), Raupach (1994) and Verhoef et al. (1997) who compared different expressions with data sets collected from large-scale international field campaigns. Following the theoretical outline and the large variability of vegetation heights, r_a is expected to vary among different wetlands of the world. E.g., a papyrus stand with vegetation height of 4 to 6 m imposes a smaller r_a, than a cattail vegetation of 0.4-0.6 m height. It is also feasible to use standard look-up tables for z_{0m} (Brutsaert 1982, Wieringa, 1986,

Monteith and Unsworth, 1990). Appendix C (Table C2) gives a review of the main biophysical characteristics measured in certain wetlands in the world.

The canopy resistance r_c for a vegetated surface is a major component of r_s. Two important analytical frameworks exist to describe r_c on the basis of the ambient conditions. Jarvis (1976) and Stewart (1988) describe r_c as a mathematical function of minimum stomatal resistance $r_{s,min}$, I_{NDV}, incoming short wave radiation R_{sd}, soil water potential ψ, (e_s-e_a) and air temperature T_a as shown by:

$$r_c = \frac{r_{s,min}}{I_{NDV}} f_1(R_{sd}) f_2(\psi) f_3(e_s - e_a) f_4(T_a)$$
(5.3)

The coefficients of the functions f_1, f_2, f_3, f_4 are determined from dedicated field measurements and/or from laboratory experiments (e.g., Stewart and Verma, 1992; Hanan and Prince, 1997). An alternative mathematical model approach to r_c is to express r_c as a function of the photosynthetical process (e.g., Jacobs, 1994; Leuning et al., 1995). This is not further elaborated here. The coupling between the conditions in the unsaturated zone and the stomatal aperture is reflected in the $f_2(\psi)$ function. Appendix C (Table C2) provides values of r_s rather than r_c, because field scale measurements reported in the appendix comprise a scale that exceeds the canopy scale.

The P-M equation can be applied to both open water and a wetland land surface type. The only difference is the value of the biophysical properties of the system. The original equation of Penman (1948) is shown in Eq. (5.4), and this version can be derived from the Penman-Monteith equation, using certain particular values of z_{0m}, z_{0h} and d. The constants are based on u_z (wind speed) expressed in km/day, and r_s is 0 s/m.

$$\lambda E_w = \frac{\Delta R_n + \gamma \left[2.7 \left(1 + \frac{u_z}{100} \right)(e_s - e_a) \right]}{(\Delta + \gamma)}$$
(5.4)

The evaporative fraction Λ is a parameter that indicates the partitioning of the available energy $(R_n - G_0)$ into H and λE, given by Eq. (4.2). The average values of all biophysical variables from the literature survey are summarized in Table 5.1. The data exhibit a wide variation, depicted by the values of the standard deviation (σ). The leaf area index I_{NDV} can be as low as 0.4 and as high as 5, which is a habitat dynamic characteristic. The albedo r_0 varies between 0.11 to 0.18. A large variation of canopy resistance is found (0 to 608 s/m) which reflects both wet and dry wetland ecosystems respectively. The roughness height z_{0m} varies between 0.009 to 0.41 m. The evaporative fraction Λ varies between 0.16 and 1.0, which corresponds to a full range of soil water availability.

The result of E_a/E_w for a range of biophysical properties is illustrated in Fig. 5.1. For this arbitrary case, r_a is calculated based on Eq. (5.2) for wind speeds from 0.5 to 8 m/s at a height of 2 m. The roughness height z_{om} of water is 0.0002 m and for wetland vegetation 0.16 m (see Table 5.1). The G_0/R_n is 0.3 and 0.1 for water and wetland respectively (assumed average values for a daily time step computation). These properties are combined with certain constants and meteorological conditions. The air temperature is taken as 27 °C and the relative humidity as 75% which implies that the vapor pressure deficit is 0.89 kPa and the slope of the saturated vapor pressure is 0.21 kPa/°C. The mean 24-hour solar radiation measured at the surface is assumed to be 250 W/m².

Table 5.1: Representative values of covered wet soil based on Appendix C (Table C2), supplemented with literature data on open water bodies and values for G_0/R_n. The aerodynamic parameters are derived from measured vegetation height

Parameter	Open water	Heterogeneous wetlands	
	Mean	Mean	σ
r_0 (-)	0.05	0.16	0.02
I_{NDV} (-)	0	2.4	1.5
E_a/E_w (-)	1.0	0.87	0.26
Λ (-)	1.0	0.73	0.24
G_0/R_n (-)	0.3	0.1	n.a.
r_s (s/m)	0	86	97
z_{0m} (m)	0.0002	0.16	0.17
z_{0h} (m)	0.00002	0.02	n.a
d (m)	0	0.90	n.a.

The P-M equation (Eq. 5.1) has been used first for covered wet soil to compute E_a. Thereafter, the values of E_w were computed with the same equation, but using the physical values for open water.

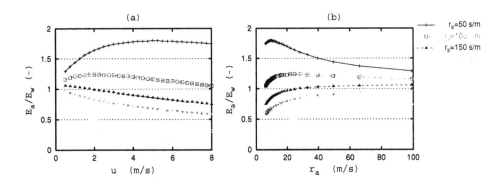

Fig. 5.1: Variability of E_a/E_w using P-M equation.

Fig. 5.1 demonstrates that – for a given climate condition – E_a/E_w can both be higher and lower than 1, depending on the possible combinations of r_a and r_s. The same wind speed blowing on a water body or on a wetland, results in different E_a/E_w depending on the value of r_s. This reveals that evaporation from a wetland cannot be determined from a single indicator such as E_a/E_w, and that instead, the biophysical system properties ought to be known. It is therefore difficult to make generic statements such as " wetlands evaporation E_a should be higher than open water evaporation E_w." (e.g., van der Weert and Kamerling, 1974), or to presume that E_a should be lower than E_w (Gavin and Agnew, 2000). This exercise teaches us that there is a wide range of characteristics that determines the evaporation from wetlands. The P-M equation is a physical-mathematical framework that is suitable for a wide spectrum of wetland ecosystems. This approach in conjunction with satellite data to update the biophysical properties monthly has been used in the present study.

5.3 Results from the Sudd

The Sudd wetland is one of the biggest swamp in Africa and belongs to the most extensive wetlands of the world. Description of the Sudd location and hydrometeorological characteristics is given is section 3.2. A brief description of the vegetation cover is given in the following section as a background for the biophysical properties interpretation presented in this chapter.

The Sudd environment supports a variety of vegetation species. Cyperus papyrus exists at the riversides and in the wettest swamps. Phragmites (reed) and Typha swamps (cattail) are extensive behind the papyrus stands and there is an abundance of submerged macrophytes in the open water bodies. Wild rice (Oryza longistaminata) and Echinochloa pyramidalis grasslands dominate the seasonally inundated floodplains. Beyond the floodplain, Hyparrhenia rufa grasslands cover the rain-fed wetlands. Acacia seyal and Balanites aegypticaca woodlands border the floodplain ecosystem (Denny, 1984; 1991). A typical toposequence of the plant species encountered in the Sudd system (close to a river channel) is provided in Fig. 5.2.

Accurate determination of the Sudd E_a is hindered by its immense size and difficult accessibility. Earlier attempts to measure E_a in the Sudd started by the experiments of Butcher (1938) and Migahid (1948) and the calculations of Hurst and Philips (1938). Chapter. 4 showed that for the year 2000 (considered as an average year), the annual rate of the Sudd E_a amounts to 1636 mm/yr, which is 20% less than assumed in earlier hydrological studies. However, since the average annual area occupied by the wetlands is 74% larger than assumed in these earlier investigations, the total evaporated volume of water is significantly more than previously expected. This means that this area still has some unresolved mysteries. The results presented hereafter are related to a fixed area that is determined on the basis of the contours of the annual (see Fig. 4.6).

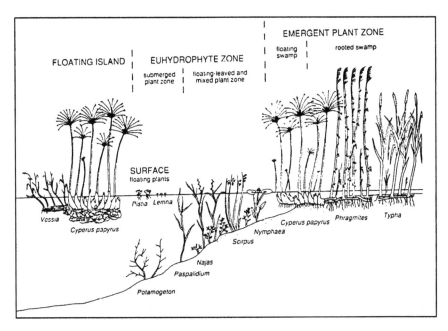

Fig 5.2: Typical toposequence of plant species in the Sudd close to a river channel (Denny, 1993).

5.3.1. Temporal variation of bio-physical properties over the Sudd

The actual evaporation of the Sudd is estimated through the application of the SEBAL remote sensing algorithm that utilizes NOAA-AVHRR images at a resolution of ~ 1 km (Chapter 4). The maximum time interval between two consecutive AVHRR images is one month. The advantage of SEBAL is that neither plant specific properties nor conditions of the atmospheric boundary layer need to be known for assessing the evaporation rates. The minimum input requirements are routine meteorological station data. The satellite image provides an excellent spatial coverage with a resolution of 1 km. The temporal coverage is limited due to cloud cover. In this study, the temporal characteristic of the Sudd E_a has been studied by repetitive calculations for 1995, 1999 and 2000. The 3 years have different hydro-meteorological characteristics in terms of rainfall and river inflow (see Table 5.2). Years with less rainfall have consistently lower inflow, outflow and a higher air temperature and vapor pressure deficit. Hence, the weather conditions markedly modify when the rainfall regime changes.

Monthly evaporation maps over the Sudd wetlands have been prepared and the related biophysical properties that cause these differences have been derived. The biophysical parameters of the Sudd were calculated from the AVHRR images using semi-empirical formulae. The parameters include surface albedo r_0, Leaf Area Index I_{NDV}, the thermal infrared emissivity ε_0 and the surface roughness z_{0m}. An example of a similar approach for the Naivasha Basin in Kenya is given in Farah and

Bastiaanssen (2001). The ground based meteorological measurements in the area for the radiation (based on sunshine duration), wind speed, temperature and vapor pressure deficit were used to compute E_a in between subsequent images. The canopy resistance r_s is calculated backward from the SEBAL E_a using the inverse P-M formula as given in Eq. (5.1).

Table 5.2: Climatic differences among the 3 years investigated.

Parameter/Year	1995	1999	2000
Precipitation mm/yr*	930	1058	950
Inflow Gm^3/yr	37.4	51.4	41.3
Outflow Gm^3/yr	16.6	18.9	17.3
Net surface radiation (W/m²)	127.3	129.3	125.8
Air temperature (°C)	28.2	27.6	27.7
Vapor Pressure Deficit (e_s-e_a) (kPa)	2.16	1.88	2.00
Wind speed (m/s)	2.56	2.58	2.64

* Average of Juba, Wau and Malakal stations

The mean values of the biophysical parameters over the Sudd derived from SEBAL averaged for the years 1995, 1999 and 2000 are tabulated in Appendix C (Table C3). The I_{NDV} shows a clear seasonality (Fig. 5.3), in accordance with rainfall season and river flooding. High I_{NDV} values > 0.50 occur during the peak rainy season Jul-Oct, and doesn't decay sharply because of the incoming Nile water, which keeps the ecosystem green for a longer period. Note that these values comprise also local habitats with a much higher I_{NDV}. This comment also applies to other system properties. Thus the absolute values on I_{NDV} are at the lower side because large portions of land are covered with open water and bare soil.

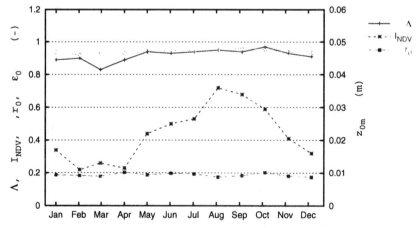

Fig. 5.3: Monthly fluctuations of I_{NDV}, Λ, r_0, ε_0 and z_{0m} over the Sudd wetland (monthly values averaged over the years 1995, 1999, 2000).

The roughness height z_{0m} follows the I_{NDV} curve (by construction) and has a peak value of ~ 0.04 m in Aug and lowest value of ~0.01 m in the dry months of Feb-Apr. The albedo r_0 behaves fairly stable in time, with a minimum value of 0.17 and a maximum of 0.20. The surface thermal infrared emissivity ε_0 varies between 0.92 in the dry season to 0.96 in the rainy season, which agrees with the literature values presented by Buettner and Kern (1965).

The distribution of the $(e_s\text{-}e_a)$ values reflects the seasonal climatology of the Sudd and its river flooding regime, Fig. 5.4. The lowest $(e_s\text{-}e_a)$ is recorded during the rainy season elapsing from Jul-Oct. The driest air coincides with the dry season in Nov-Apr. As affected by river flooding and upwind atmospheric boundary conditions, the $(e_s\text{-}e_a)$ in Nov-Dec remains lower than in Feb-Apr. The net surface radiation R_n reflects variation of the solar radiation, the cloud cover and the net long wave radiation. Due to the higher cloud coverage during periods of higher solar radiation in Aug-Oct, the net radiation remains fairly stable with time. This has an important impact on E_a, which usually obeys the temporal pattern of R_n. The slope of the vapor pressure curve $\varDelta$ shows small variability throughout the seasons (0.18 to 0.24 kPa/°C) because the temperature behaves fairly stable throughout the year.

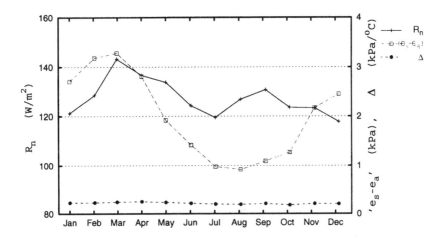

Fig. 5.4: Monthly fluctuations of climate parameters R_n, $(e_s\text{-}e_a)$ and $\varDelta$ over the Sudd wetland (monthly values averaged over the years 1995, 1999, 2000)

The hydrological control on evaporation depends on moisture availability in the root zone, which governs the surface resistance r_s. Fig. 5.5 shows that the surface resistance r_s has a distinct seasonal variability, which is consistent with the inter-seasonal variation of the $(e_s\text{-}e_a)$ and I_{NDV} as formulated in Eq. (5.3), as well as the river flow regime and the related ground water table (GWT) fluctuations. Computation of GWT is discussed in section 5.3.3. The lowest r_s values are associated with the lowest $(e_s\text{-}e_a)$ and the highest I_{NDV} during the wet months Jul–Oct, and the reverse occurs during the dry months Feb-Apr with r_s>300 s/m. The aerodynamic resistance r_a appears to be much smaller than r_s, thus r_s dictates the total resistance for water vapor transport for most of the year. During Jul–Oct, the values of r_s and r_a are similar. It should be realized that Fig. 5.5 is based on manifold SEBAL computations that are based on satellite observations and weather data only. This energy balance is independent of the water balance (section 5.3.3), which also supports the depicted seasonality of r_s.

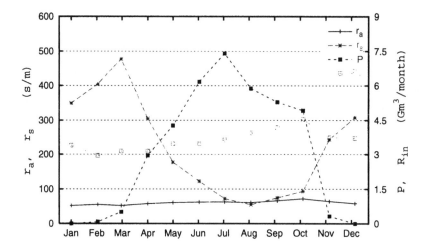

Fig. 5.5: Monthly fluctuations of surface r_s, aerodynamic r_a resistances, inflow R_{in} and rainfall P over the Sudd wetland (monthly values averaged over the years 1995, 1999, 2000)

The spatial variation of r_s is further elaborated in Fig. 5.6, which shows the frequency distribution for a wet month (Aug) and a dry month (Mar). More pixels have low r_s values during the wet month. The median for the three years in Fig. 5.6 (Mar) lays approximately between 350 to 375 s/m. It is to be noted that wetness over the Sudd in March is mainly provided by the river water supply (negligible rainfall). There are small differences of r_s between the years, but still compatible with the river water supply. Fig. 5.6 reflects a Sudd that has insufficient moisture in Mar to keep all vegetation green. Hence, large portions of intermediately wet soil are present in this data set.

During the wet season (Aug), r_s reduces to very low values. In a wet year, the median is 0 s/m only, which coincides with the lower end of the data found in Appendix C (Table C2) and suggests that in these areas water is ponding at the surface. In relatively dry years; 2000 and 1995, the median is 50 s/m and 100 s/m respectively, which means that soil is wet but that the majority of the area is not ponded.

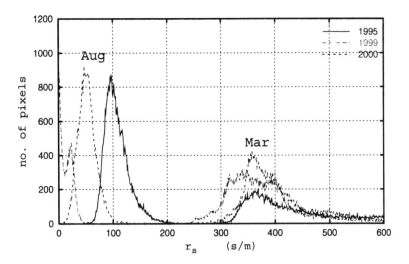

Fig. 5.6. Frequency distribution of surface resistance r_s over the Sudd during the dry month of Mar and during the wet month of Aug (1 pixel ~ km^2).

The biophysical properties of worldwide wetlands r_0, I_{NDV}, E_d/E_w, Λ, r_s and z_{0m} reviewed in Appendices C (Table C1 and Table C2) are compared to the values derived from remote sensing over the Sudd wetland in Fig. 5.7. Parameter values (mean±σ) normalized by the literature mean (Table 5.1) are compared to the Sudd means for 1995, 1999 and 2000. Furthermore, the corresponding mean values of wet season Jul-Oct and dry season Jan-Apr are plotted in the figure.

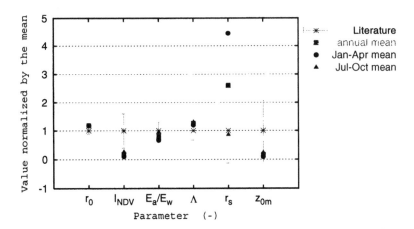

Fig 5.7: Comparison of the biophysical properties derived from literature and from SEBAL over the Sudd. Error bars represent one σ normalized by the mean.

A wide variation is depicted for I_{NDV}, r_s and z_{0m}. The value of r_0, E_a/E_w, and Λ show smaller variations. The average E_a/E_w value is 0.70, which falls within the range quoted in Appendix C (Table C1). It can be concluded that the biophysical parameters in the summer period Jul-Oct are all within the expected range, and indeed, close to the mean value of the literature. The winter values Jan-Apr for the Sudd lay more on the edge. This shows that the Sudd in the winter does not behave as a real wetland.

5.3.2.   Temporal variation of evaporation in the Sudd

The monthly values of the actual evaporation E_a and the ratio E_a/E_w over the Sudd for the three years 1995, 1999 and 2000 are given in Appendix C (Table C4). E_w is the open water evaporation that is been computed by means of Eq. (5.4) to allow comparison with the data given in Appendix C (Table C1). G_0/R_n assumed negligible both for wetland and water surface in a monthly time step. The result of E_a/E_w in Fig. 5.8 is compatible with the one given in Fig. 5.1, i.e., similar result is obtained if the climate condition of the Sudd is applied to the theoretical example of Fig. 5.1.

The differences of E_a can be large between the years; 13% between 2000 (1643 mm/yr) and 1995 (1460 mm/yr), and up to 33% between 1999 (1935 mm/yr) and 1995 (1460 mm/yr). Rainfall and the size of the flooded area are positively correlated to the annual E_a.

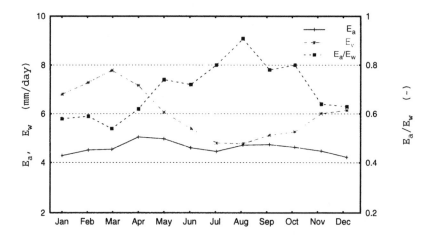

Fig 5.8: Monthly fluctuation of the Sudd E_a and open water E_w (monthly values averaged over the years 1995, 1999, 2000).

While E_w shows clear seasonality in accordance with (e_s-e_a) (see Fig. 5.4), E_a is extremely stable due to compensating effects. The possible explanation for the quasi-steady variation of E_a in contrast with E_w, is that net radiation varies only between 120 to 150 W/m² and that (e_s-e_a) and r_s have cancelling effects due to their natural feedback mechanisms as described by Jarvis (1976) and Stewart (1988). This is an important conclusion for this tropical watershed in Sudan.

Fig. 5.9 depicts the spatial distribution of the mean E_a for the 3 years 1995, 1999 and 2000. The catchment boundary is defined according to an E_a threshold of 1550 mm/yr (3.85×10^6 ha). Fig. 5.9 shows that highest annual evaporation occurs at the permanent swamps close to the river channels, while the seasonal swamps evaporate relatively less, but still much higher than the surrounding lands.

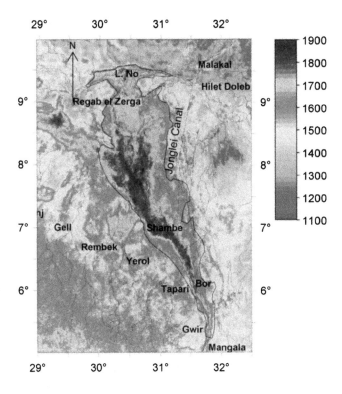

Fig. 5.9: Contour lines of average annual E_a (data presents annual average for the years 1995, 1999, 2000).

5.3.3. Qualitative description of soil moisture temporal variability

One factor, often overlooked in the Sudd hydrology, and perhaps of importance to explain the sharp seasonality of the surface resistance, is the intra-seasonal availability of moisture in the non-inundated areas within the Sudd. The Sudd wetland is composed of permanent swamps, close to the river course, and seasonal swamps created by river flooding and rainfall spells. The boundaries of both types of swamps are not accurately defined, and vary with time. The availability of water in the unsaturated zone is an important source of moisture outside the rainy season. A crude calculation of the amount of water within the root zone demonstrates that the high surface resistance over the Sudd outside the rainy season between Nov and Apr can be attributed to a dry soil moisture condition, which is related to a lower ground water table (GWT).

The monthly variation of the GWT has been calculated based on the Sudd water balance for the years 1995, 1999 and 2000. The river inflow computed at Mangala R_{in} is estimated based on Victoria outflows (corrected for Lake Kayoga and Lake Albert contributions) and the rainfall at Juba. The latter is used to estimate the torrents flow between Lake Albert and Mangala. Monthly rainfall P over the Sudd is taken as the average of Juba, Malakal and Wau. Outflow R_{out} has been estimated by the method of Howell et al. (1988). Evaporation E_a is estimated by SEBAL over the Sudd area ($A=3.85 \times 10^6$ ha). The monthly change of storage volume dS/dt within the Sudd is computed by Eq. 4.7. It is assumed that the Sudd wetlands is composed of wet surfaces (reservoirs), and non-inundated areas with the GWT below the surface (unsaturated zone). Therefore, the total dS/dt can be decomposed into two components: dS_w/dt change of storage volume in the open water parts and dS_{unsat}/dt change of storage volume in the non-inundated parts (Eq. 5.5):

$$\frac{\mathrm{d}S}{\mathrm{d}t} = \frac{\mathrm{d}S_w}{\mathrm{d}t} + \frac{\mathrm{d}S_{unsat}}{\mathrm{d}t} \qquad (5.5)$$

The soil moisture in the topsoil layer is estimated from the satellite images as a function of the evaporative fraction (Eq. 4.3). For simplicity, it is assumed that the average soil moisture over the entire Sudd wetland applies to unsaturated soil conditions. This is strictly seen, not correct, but may allow a qualitative description of the GWT temporal variability. First, the change of storage volume dS_w/dt and the associated area A_w of the open water parts within the Sudd are estimated based on the hydrological model of Sutcliffe and Parks (1999), given by Eq. 4.6. In this, model the relation between A_w and S_w is schematized to be linearly related as $A_w=kS_w$, with a constant depth $1/k= 1$ m. The corresponding E_w is calculated based on P-M applied to a water body using the meteorological data over the Sudd.

Using dS_{unsat}/dt, and the soil moisture information over A_{unsat}, the change of moisture storage depth in the unsaturated zone can be estimated. If it is assumed that saturated soil moisture condition within the Sudd is associated with a shallow groundwater table that is close to the surface, then it is possible to describe – at least qualitatively – the seasonal variability of the GWT. An initial GWT is selected for 1995 and 1999 to attain a maximum level just at the land surface. It is assumed that the GWT for the entire Sudd is at the surface in Oct. Appendix C (Table C5) gives the results for each of the 3 years, and Fig. 5.10 shows the average distribution.

Since these computations, involve several assumptions, only some qualitative conclusions can be drawn. The variability of the GWT in the Sudd and its implications is coherent with earlier results derived independently on the evaporation E_a and biophysical properties. The GWT drops to the lowest level in May just before the rainy season and continues rising after the rainy season, reaching its peak in Oct, which is linked to the peak river inflow. The surface resistance variability from remote sensing technology is clearly negatively correlated. Comparison of Fig. 5.10 to Fig. 5.9 indicates that substantial parts of the Sudd wetlands – which achieve substantial amounts of E_a – are probably not covered permanently with water. This important conclusion indicates that the Sudd is covered with open water only near the riverbed. The adjacent wetland environment undergoes a seasonal flooding only.

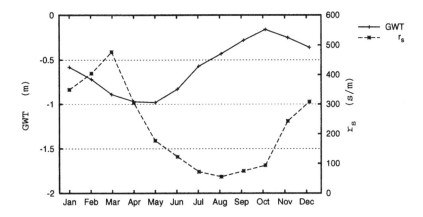

Fig. 5.10: Monthly variation of the aerial average GWT in the un-inundated parts (monthly values averaged over the years 1995, 1999, 2000).

5.4 Conclusions

The Sudd evaporation E_a doesn't show much seasonal variability, while open water evaporation E_w clearly follows the seasonal climatic variation. This quasi-steady state of E_a in the Sudd is attributed to the lack of seasonality of net radiation and the canceling effect of the vapor pressure deficit and surface resistance throughout the season. The distinct seasonality of the surface resistance over the Sudd could be explained by the vapor pressure deficit and soil moisture feedbacks on stomatal aperture. The variation due to rainfall and upstream Nile inflow is responsible for the seasonal fluctuation of the groundwater level. The Sudd wetland has a profound seasonal variation of the groundwater table, which demonstrates that a majority of the Sudd is non-inundated and has a seasonal decaying vegetation system. Only the parts closest to the riverbed are permanently saturated. The lower groundwater table reduces the soil moisture in the root zone and increases the leaf water potential, which induces water stress.

Consequently, the concept of E_a being proportional to E_w does not hold true for the marshlands of the Sudd, which implies that the evaporative depletion from the Sudd needs to be re-examined. Appropriate attention has to be given to other wetlands in monsoonal climate systems also, that may have a similar seasonality, e.g., the Okavango. Although this may have been known before, this is among one of the first attempts to quantify this seasonality and demonstrate that E_a is not proportional to E_w.

The difference in rainfall between dry (930 mm/yr) and wet years (1058 mm/yr) is only 14 %. However, the range of evaporation between dry years (1460 mm/yr) and wet years (1935 mm/yr) exceeds by 33% the differences in rainfall, because E_a originates from both rainfall and inflow. This implies that evaporative depletion of the Sudd is very sensitive to both local and upstream rainfall patterns.

The Penman-Monteith equation appeared to provide a solid physical basis, but the spatial distributed biophysical properties must be known. It is demonstrated in this chapter that these properties can be derived from remote sensing techniques, in particular the surface resistance to evaporation that is the most complex parameter of a heterogeneous vegetation system to determine. The biophysical properties agree – especially during the summer season – with the values measured elsewhere.

6. Regional climate modeling of the Nile Basin[1]

6.1 Introduction

It is increasingly recognized that appropriate water resources planning and management at a river basin level is viable only by considering the complete water cycle in the basin, i.e., including both the land surface (hydrological) and the atmospheric processes. In many river basins, steady climatic conditions are no longer considered a valid assumption for sustainable water resources management. Therefore, despite its significant computational effort water resources studies at river basin level are increasingly linked to regional climate studies. Examples are studies on impact of climate change/variability on water resources or the studies of land use change and their impacts on regional climate and water resources (see a review in Watson et al., 2001). Similarly, it is appreciated that adequate representation of the land surface in climate models is crucial for accurate modeling results (Kite, 1998; Koster et al., 2002).

The Nile Basin experiences rising demands for its (limited) water resources, due to among others, increasing population growth and rising economical development. As a result, there is increasing pressure to augment river discharge by reducing the non-beneficial evaporation losses from the Upper Nile swamps and from shallow water table areas (Sebeka's or playas). In the Sudd, the Nile evaporates more than half of its local flow (see section 3.2). Similarly, a substantial amount of water evaporates from the neighboring Bahr el Ghazal and the Sobat swamps. The total amount of evaporation from these 3 adjoining upland watersheds is around 45 Gm^3/yr, which is approximately half the annual inflow into Lake Nasser (PJTC, 1961). The proposed approach for water saving is to build river short-cut channels to prevent spillage into the swamps and divert the flows downstream into the main channel (e.g., the uncompleted Jonglei canal).

Although these plans for the Nile Swamps were initiated about 100 years ago, and associated with intensive environmental impact assessment studies (see a review in Howell et al., 1988), no genuine attempts have been made to study the impact on the regional climate. Regional climate model (RCM) simulations can be instrumental to evaluate the impact (if any) of these large-scale land use modifications. A coupled RCM would enhance the understanding of the complete water cycle and the imbedded feedbacks, allowing a more integrated approach of water resources planning in response to the critical water shortage in the Nile.

---

[1] Based on: Mohamed, Y.A., van den Hurk, B.J.J.M., Savenije, H.H.G., Bastiaanssen, W.G.M. 2004. Hydroclimatology of the Nile: Results from a regional climate model, submitted to HESS and published in HESSD.

.

In the absence of a good understanding of the Nile water cycle, different and sometimes contradictory conclusions were drawn on the impact of the Sudd wetlands on the climate regime of the Nile. The JIT (1954) and Howell et al. (1988, p. 375) suggested no impact on the regional climate by draining part of the Sudd (Jonglei canal). Eltahir (1989), Eagleson (1986) among others hypothesized that the evaporation from the Sudd would be felt climatically over a wider region. The discussion on the Sudd reclamation remains unresolved despite the importance of the wetland to the ecosystem and local inhabitants on the one hand, and the potential it possesses for additional water supply on the other hand.

Worldwide, RCM's are used for a variety of applications related to the hydrology and water resources of river basins (see e.g., a review of Giorgi and Mearns, 1999). Bonan (1995) and Sun et al. (1999a) used RCM's to study the influence of the Nile source water (Equatorial lakes) on the regional climate. They demonstrated a strong atmosphere-lake interaction that significantly modulates the regional climate of East Africa. Sun et al. (1999b) also showed that there is a strong positive correlation between the Upper Nile precipitation (over lake Victoria) and the warm El Nino-Southern Oscillations ENSO. This is also confirmed by Farmer (1988), Nicholson (1996) and others. Results from GCM were used to study the impact of climate change/variability on the Nile water resources, e.g., Conway and Hulme (1996). The IPCC Third Assessment Report, Working Group II (Watson et al., 2001) gives a review of the possible impacts of climate change on the Nile water resources. The report shows the difficulty in predicting the Nile response to global warming because of the fact that different simulations give conflicting results. Unlike the Amazon and the Mississippi basins, no RCM study has been made to investigate the impact of land use changes on the Nile climate.

In the present study, the Regional Atmospheric Climate MOdel RACMO (Lenderink et al., 2003) is run over the Nile for the period 1995 to 2000. The objective is to obtain a better understanding of the water cycle over the Nile and the embedded feedbacks between land and atmosphere. The model is forced by the ERA-40 (ECMWF Re-analysis 1957-2001) boundary condition, and adjusted to simulate the routing of the Nile flow over the Sudd swamps. Model evaluation is based on observational data sets of various sources: radiation, precipitation, evaporation and runoff.

The validated model has been used to compute the moisture recycling over the basin as defined in section 2.3. Earlier results made by a GCM show a considerable precipitation recycling over East Africa including the Nile Basin. By using particle tracers in GCM simulations on an 8°x10° model grid, Koster et al. (1986) claimed a significant contribution of the evaporated water from the Sudd to the regional rainfall. Using the same methodology, but with a finer GCM, Bosilovich et al. (2002) showed that a substantial percentage of precipitation over the Nile basin is originated from land evaporation (could be outside the basin). The outcome of these results is limited due to the course model resolution that misses the details of the Upper Nile swamps.

This chapter presents and discusses the results of the Nile climate model with particular emphasis on the hydrology of the upstream wetlands, and the moisture recycling ratios obtained. The same model is applied to a land use change scenario over the Nile (drained wetlands) presented in Chapter 7. Please refer to Chapter 3, for description of the basic features of the Nile hydroclimatology.

6.2 The Regional climate model

6.2.1.   RACMO basic features

The Regional Atmospheric Climate Model (RACMO) is based on the HIRLAM model (HIgh Resolution Limited Area Model) in combination with the physical parameterization from the ECMWF model (Lenderink et al., 2003). It is the main limited area model used by KNMI (The Royal Netherlands Meteorological Institute) for climate research. RACMO has been applied to the Nile domain between 12°S, 35.96°N and 10°E to 54.44°E (see Fig. 6.1). The model grid is 0.44x0.44° ~ 50 km x 50 km resolution. The total number of grids is 102x110 = 11220. Vertically, the model is divided into 31 layers (hybrid coordinate system). The time step is 10 min, and the simulated period covers 1995 to 2000.

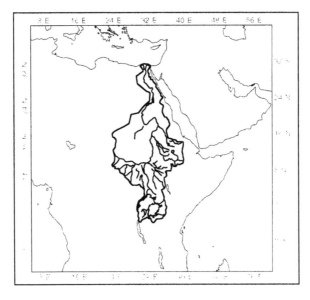

Fig. 6.1: Location of model boundary enclosing the Nile Basin.

The initial atmospheric fields and lateral boundary conditions were interpolated from the ERA-40 data. The vegetation cover was retrieved from the GLCC Global land Coverage Characteristics data set as classified to the ECMWF land surface scheme. The remaining surface parameters (geo-potential height, orographic variability) were interpolated from the HIRLAM climate system. The land surface scheme of RACMO is based on the so-called Tiled ECMWF Scheme for Surface Exchanges over Land (TESSEL; (van den Hurk et al., 2000)). Each land grid box is composed of 6 tiles representing various fractions of bare ground and vegetation. The soil below the surface is composed of 4 layers with fixed depths being 0.07, 0.21, 0.72 and 1.89 m thickness, respectively. The soil physical properties are uniform for all model grids. The precipitation on a grid box is partitioned into interception and throughfall. The interception is a function of the type of rain (convective, large scale) and the Leaf Area Index. The throughfall infiltrates into the soil, where vertical water exchange obeys Darcy's law through the 4 soil layers. Turbulent fluxes (sensible and latent heat) are computed based on resistance parameterisation of the respective tiles (open water, bare soil, vegetative cover) that represent surface and aerodynamic properties and the soil moisture conditions. The remaining net heat flux is transferred to the soil. Surface runoff occurs when the throughfall exceeds the infiltration capacity (this rarely occurs). The main runoff component is the deep runoff through the bottom of the 4th soil layer (i.e., free drainage). The original RACMO model doesn't provide routing of runoff to the catchment outlet.

6.2.2. Model adjustments

Several 1-year model runs were performed to define the necessary adjustment of RACMO to the Nile conditions. First, inspection of the ECMWF physics based on the GLCC maps shows that the Sudd, and some areas farther north have been classified as 'bogs', to which a very high minimum canopy resistance ($r_{s,\ min}$ = 240 s/m) is assigned (van den Hurk et al., 2000). This is unrealistic for the typical wetlands vegetation over the Sudd (e.g., Lafleur and Rouse, 1988). Results were improved after replacing the land cover characteristics of the GLCC with ECOCLIMAP (Masson et al., 2003). The canopy resistance $r_{s,\ min}$ over the Sudd is reduced to 15 s/m to mimic the wetland evaporation characteristics. The evaporation results were also improved when using deeper soil layers at: 0.07, 0.33, 1.27 and 3.32m following Lenderink et al. (2003) so that a total soil depth of 5 meter is formed.

A specific characteristic in the Nile Basin is the wide spreading of the river flow over the Sudd swamps. Since RACMO doesn't include the runoff routing process, this had to be introduced by explicitly transferring the runoff from the upstream catchment to the Sudd. Every day during the model simulation, an additional amount of water stemming from the upstream runoff is distributed equally over the 15 grid points of the Sudd. A spin up time of 1 year is used to approach a realistic initial soil moisture condition.

Evaluation of the radiation results against field measurements (from two gauging stations) showed that the default radiation parameterization underestimates the incoming short wave radiation, while it computes realistic incoming long wave radiation. This could be corrected by adjusting the amount of aerosols. In addition, over the Ethiopian highlands, the model originally computed unrealistically high precipitation. After smoothing the orography, this was substantially improved. Unrealistically high river runoff was obtained by the default drainage coefficients, and reasonable estimates were obtained when the saturated hydraulic conductivity was reduced by a factor 10 to represent better the character of flooded alluvial soils. The results presented hereafter are based on these modifications.

6.3 Observational data used for model calibration

The observational data sets used for assessing model output include: ground stations of meteorological data, satellite derived estimates of evaporation, precipitation and river discharge data.

6.3.1. River discharge

Reliable discharge measurements are available at 11 key locations along the Nile River system (see Fig. 3.3). Except for the sub-catchments upstream of Malakal, all tributaries of the Nile are gauged. The water balance of the different river reaches allows inspection of the flow time series and estimation of the irrigation abstractions and/or the evaporation from storage reservoirs. The climate model computes natural river flows as the sum of all upstream free drainage fluxes. The measured gauged flows were corrected for withdrawals and evaporation losses occurring upstream of the gauging stations. The resulting river flows at the stations Khartoum (Blue Nile), Khartoum (White Nile), Khartoum (Main Nile), Atbara (Atbara River) and Dongola (Main Nile) are the natural river flows at the outlet of the sub-basins. The travel time of the flood wave from the source at the Ethiopian Plateau to Aswan is around 2 to 3 weeks during the flood season (Jul to Oct), while the travel time along the White Nile is much longer, around 8 weeks from lake Victoria to Aswan. However, the flow of the White Nile is relatively steady. Therefore, for comparison between model results and observations on a monthly time scale, no correction for travel time was necessary.

6.3.2. Precipitation

Four sources of precipitation data have been considered: (i) The Sudan meteorological department ground stations data, (ii) The Global Precipitation Climatology Center (GPCC), providing monthly rainfall data at 1° resolution, interpolated from conventional gauged observations (Rudolf et al., 2003), (iii) The Famine Early Warning System (FEWS), providing 10-daily rainfall data at 0.1° resolution, based on METEOSAT 5 satellite data, gauged data and modeled data (Herman, et al., 1997), (iv) The MIRA (Microwave Infrared Algorithm) data combining satellite passive microwave and infrared data to produce daily mean precipitation at 0.1° resolution (Todd et al., 2001). Fig. 6.2 shows the locations of GPCC and the Sudan stations. No station data were available from the other riparian countries.

A rigorous inspection has been done for the daily data of the Sudan gauging stations, which were aggregated to monthly values. This data set is considered as reference for the comparison with the other data sets.

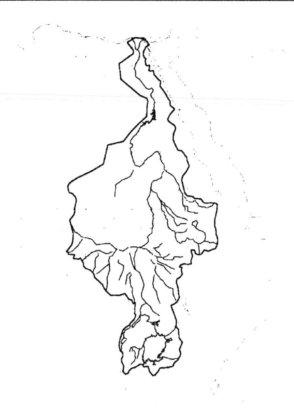

Fig. 6.2: Location of the precipitation stations (+ GPCC, ▢ Sudan gauging
stations)

The average values for the 30 point locations of the Sudan stations are presented in
Fig. 6.3. Note that GPCC and Sudan stations are supposed to be from the same
source, i.e., gauged data of the Sudan meteorological department. The two curves
are not identical, but indeed closest, except in 1996. Differences may be attributed to
the fact that the GPCC data pass an automatic quality control. No further details are
available on how the GPCC data has been corrected and averaged. The MIRA data
is generally about 50 % higher than all 3 data sets, probably due to the inclusion of
the radar data. Except for a few months, the FEWS data set (as expected) is close to
GPCC. The GPCC data set is used for the evaluation of model results, owing to its
fair comparability to the reference data set (except in 1996) and the spatial extent of
the data.

6.3.3.   *Evaporation*

The hydrometeorological observations over the Upper Nile swamps are very scarce
(the area has been a war zone since 1983). Remote sensing based estimates can be
instrumental to fill in the gaps of hydrological knowledge. Therefore, NOAA-

AVHRR LAC images were acquired over an area of 1000 km x 1000 km, covering the swamps of the Sudd, the Bahr el Ghazal and the Sobat basins (see Fig. 3.3). Monthly evaporation maps were derived using the SEBAL algorithm. SEBAL is a parameterization scheme of the surface heat fluxes based on spectral satellite measurements. Monthly (actual) evaporation, evaporative fraction and soil moisture maps were prepared for 3 years of different hydrometeorological conditions 1995, 1999 and 2000. Please refer to section 4.2 for further details on SEBAL and the derived evaporation data. The evaporation computed from SEBAL has been validated by checking water balance computations of 3 sub-basins: Sudd, Bahr el Ghazal downstream discharge stations and Sobat (section 4.4). Acceptable results were obtained for the Sudd and Sobat, while the balance doesn't close for the Ghazal Basin. This is due to the underestimated surface inflow to this swamp. There is no information to evaluate the SEBAL evaporation for the areas outside these 3 sub-basins.

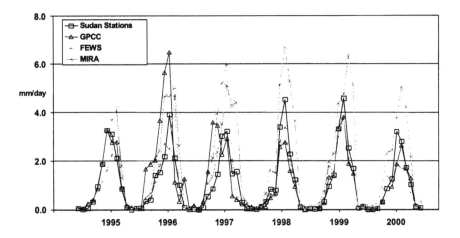

Fig. 6.3: comparison of precipitation from 4 sources: Sudan stations, GPCC, MIRA and FEWS; (mean of the 30 point locations of the Sudan stations).

6.3.4. Radiation

Only few radiation measurements are available within the model domain. Sunshine duration is available for the Sudan stations, which is routinely used to calculate solar radiation. Observed radiation data at two stations could be acquired: Riyadh (Saudia Arabia) at 24.7°N, 46.7°E and Ndabibi (Kenya) at 0.5°S, 36.2°E. They are located outside the basin, but within the model domain. The Riyadh data are archived at the World Radiation Monitoring Center WRMC, the data center of the Baseline Surface Radiation Network BSRN, available at http://bsrn.ethz.ch/. At the Ndabibi station, measurements of the incoming short wave radiation for 1998 at 20-min interval are available (Farah, 2001).

6.4 Model results and discussion

Since the sub-basins of the Nile have different physical and hydroclimatological characteristics, it is noteworthy to evaluate model results over both the whole Nile basin (Main Nile) and the sub-basins separately. Monthly time series results (1995 to 2000), or mean annual cycles are presented for the Sudd basin, White Nile, Blue Nile, Atbara River and the Main Nile, locations of sub-basins is shown in Fig. 3.3. The results presented constitute: runoff, precipitation, evaporation, soil moisture storage, radiation and moisture recycling.

6.4.1. Runoff

Model runoff R is compared to the river discharge gauged at catchment outlets of the 4 sub-basins Atbara, White Nile, Blue Nile and the Main Nile (see Fig. 6.4). The location of the sub-basins and discharge measuring stations is given in Fig. 3.3. Values are expressed in m^3/s, to allow inspection of the relative contribution of each sub-basin to the total flow at Aswan. It is to be noted that RACMO computation doesn't include groundwater flow below the 5 m depth, i.e., the calculated R may includes a groundwater recharge term, which believed to be small and can be neglected.

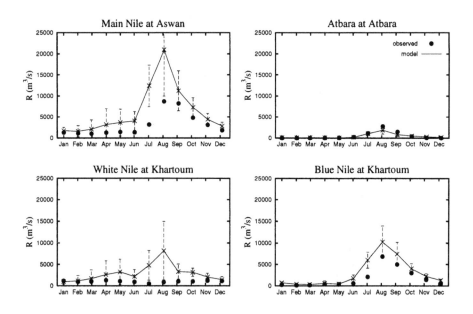

Fig. 6.4: Monthly model and observed runoff R (mean annual cycle 1995 to 2000). Error bars indicate one standard deviation (σ) of the monthly means.

The model overestimates R over the White Nile, and hence on the Main Nile. The results over the Ethiopian Plateau (Atbara and Blue Nile) are reasonably well in magnitude and time evolution. It is to be noted that a small error, e.g., in precipitation and/or evaporation over the White Nile can results in an excessive error of runoff because of its extremely low runoff coefficient. E.g., an error of P of 0.2 mm/day produces an error of about 4,000 m³/s in runoff. The mean annual runoff coefficient R/P of observed P and R in 1995 to 2000 for the Main Nile, Atbara, White Nile and the Blue Nile are 0.05, 0.16, 0.02 and 0.19, respectively, and the corresponding results derived from the model are 0.14, 0.17, 0.09 and 0.29. While tuning the model, we were more inclined to obtain optimal results over the Atbara and Blue Nile catchments rather than the White Nile since about 5/7 of the Nile runoff is generated from these two catchments.

6.4.2.   Precipitation

The comparison of the mean model precipitation P against observations of the GPCC is given in Fig. 6.5. On the Ethiopian catchment (Atbara and Blue Nile) – where most of the Nile runoff is generated – the model produced reasonable results, except for the rain peak on the Blue Nile, where the model slightly overestimates precipitation. The model accurately captures seasonality of the rains over the 4 sub-basins. On the White Nile the model underestimates the rain during the Mar to Jun period.

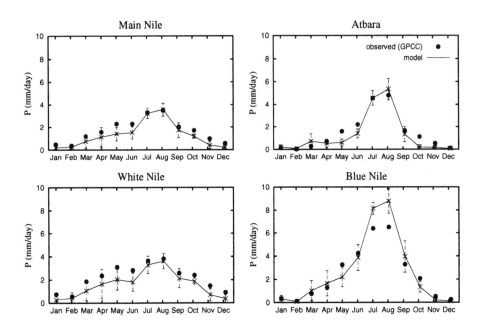

Fig. 6.5: Monthly model and observed (GPCC) precipitation P (mean annual cycle 1995 to 2000). Error bars indicate one σ of the monthly means.

Further inspection of the seasonal precipitation during a sample year of 1999, May, Apr and May (MAM) over the White Nile reveals that the underestimation is also present in the ERA-40 data. The RACMO bias is likely related to the lateral forcing imposed on the southern boundary of the model (Fig. 6.6). The ERA-40 places the MAM precipitation more south of the Bahr el Ghazal Basin than the GPCC data. RACMO computes compatible results to the reanalysis data, although the negative bias seems to be a bit more pronounced.

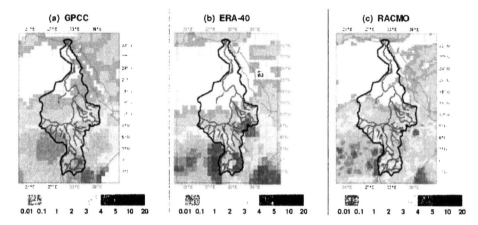

Fig. 6.6. Map results of Mar to May 1999 precipitation in mm/day GPCC, ERA-40 and RACMO model.

6.4.3.   Sudd water balance components

To simulate the inundation of the Sudd by Nile water, a constant inflow R_{in}=4.1 mm/day is distributed every day over the 15 grid points of the Sudd. The 4.1 mm/day is the observed runoff from the Nile catchment upstream the Sudd, and it is a major evaporation source in the dry winter season. Fig. 6.7 shows a closed water balance derived from model results over the Sudd, where Bal. = $P+R_{in}-R_{out}-E-\mathrm{d}S/\mathrm{d}t$.

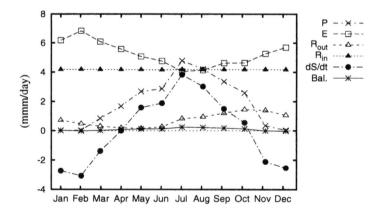

Fig. 6.7: Hydrological balance over the Sudd wetland in mm/day computed with the calibrated RACMO model (mean annual cycle 1995 to 2000)

The term dS/dt is the change of sub-surface water storage (soil moisture) in the 4 soil layers. The data represent the mean annual cycle for the 6 years 1995 to 2000.

The comparison of model results to observations P, E, R_{out} and dS/dt is given in Fig. 6.8. In this case the model dS/dt is computed for the top 2 layers (0.4 m thick), to allow comparison with dS/dt estimated from remote sensing by SEBAL .The data represent a mean annual cycle for the 3 years: 1995, 1999 and 2000. In general, except for the dry months, the model reproduces P, E and dS/dt fairly well. E is overestimated during the dry months Nov to Apr, but closely resembles remote sensing data during the wet season May to Oct. The model reasonably reproduces the variability of the soil moisture storage dS/dt. The model underestimates the outflow runoff R_{out} by about 1.5 mm/day (~600 m^3/s). The mismatch of the flow discharge from the Sudd area is small compared to the White Nile flow presented in Fig. 6.4.

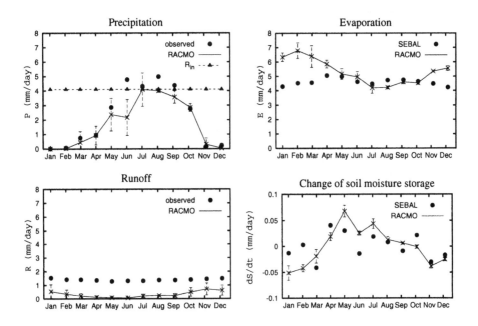

Fig. 6.8: Precipitation, Evaporation, Runoff and Soil moisture variation over the Sudd wetland (mean annual cycle: 1995, 1999, 2000). Error bars indicate one σ of the monthly means.

It appears that there is a clear seasonality in the model E over the Sudd in response to the available energy and atmospheric demand (higher during the drier months), whereas the remote sensing data show a quasi-steady evaporation. This can partly be attributed to the seasonality of the surface resistance r_s to evaporation. The Sudd system is now parameterized as one large floodplain, but during the dry season, not all land is flooded. Hence, land at little higher elevation dries out, which boosts up the r_s and reduces the E flux, see section 5.3 for further details on the temporal variability of the Sudd evaporation and biophysical properties. The model assumes constant leaf area index I_{NDV} throughout the year, so it doesn't adjust the r_s during the dry season (low I_{NDV}), while SEBAL accounts for variability of r_s with I_{NDV}. Considering the objective of the modeling study, to investigate the impact of the Sudd wetland on the Nile hydroclimatology, in particular during the rainy season, it is considered that model results are satisfactory.

6.4.4.   Radiation

Incoming short wave R_{sd} and long wave R_{ld} radiation at the land surface observed at Riyadh and incoming short wave radiation R_{sd} at Ndabibi are compared to model results in Fig. 6.9.

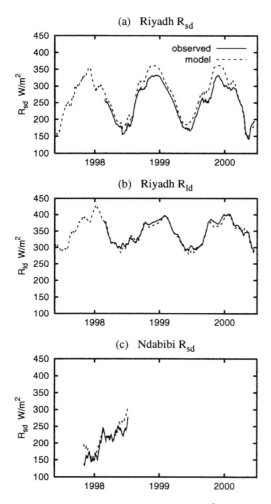

Fig. 6.9: Radiation results in W/m^2: (a) Incoming short wave radiation at Riyadh, (b) Incoming long wave radiation at Riyadh, (c) Incoming short wave radiation at Ndabibi (30 days moving average).

In the original formulation of RACMO, R_{sd} in Riyadh was underestimated by 20 to 40 W/m^2 (~10 to 20%). To remove this bias, the climatological aerosols content was reduced, but from Fig. 6.9 it seems that the aerosol reduction has been slightly too strong. Note the difference in seasonal phasing of R_{sd} between Riyadh at 24.7°N and Ndabibi at 0.5°S. Peak R_{sd} at Riyadh occurs during the northern hemisphere summer associated with the lowest R_{sd} at Ndabibi. The only available long wave radiation measurements at Riyadh shows that RACMO could reproduce R_{ld} quite accurately.

6.4.5.   Total Nile Basin water cycle

The time series of the regional water cycle components; Q_{in}, Q_{out}, P, E, R, dS/dt over the Nile area are given in Fig. 6.10, and in tabular format in Appendix D (Table D1). The data comprise the mean annual cycle of model results during 1995 to 2000 averaged over the whole Nile area. P, E and R are based on 6 hourly data, Q_{in} and Q_{out} are based on 12 hourly, and dS/dt are based on daily data. The annual cycle of the fluxes is not as pronounced as for the smaller sub-catchments, however, net convergence occurs during Jun, Jul, Aug and Sep, and divergence in Dec, Jan, Feb, Mar. Obviously, P, E and R are higher during convergence months, and reduced during divergence time. Sub-surface storage (within the 4 soil layers) occurs during the rainy months, and depleted during the dry months, resulting into a zero annual mean.

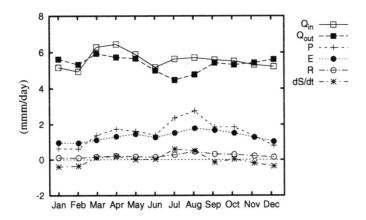

Fig. 6.10: The components of the Nile regional water cycle in mm/day (annual mean 1995 to 2000)

As discussed in section 3.1.1, the Nile water cycle has both the characteristics of a single rainy season during Jun to Sep (Ethiopian Plateau; Blue Nile and Atbara), and that of a double rainy season (Equatorial Lakes Plateau, part of the White Nile). This is clearly depicted by Fig. 6.11, which shows the spatial distribution of the atmospheric horizontal water transport (arrows) and the water vapor convergence/divergence in the two distinct seasons.

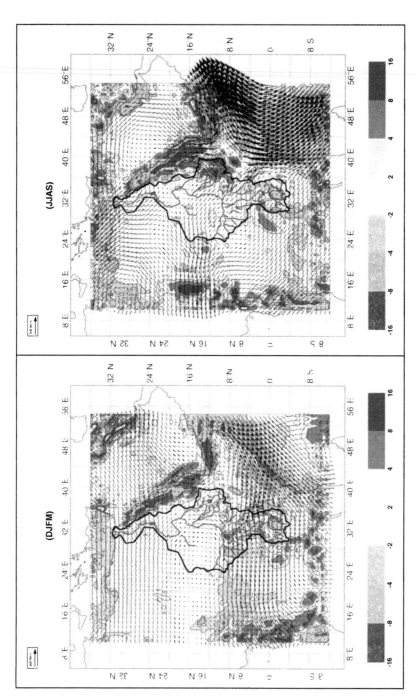

Fig. 6.11: Spatial distribution of atmospheric moisture fluxes for the two main seasons (>0 is convergence, <0 is divergence in mm/day). DJFM = Dec to Mar, JJAS = Jun to Sep.

Over the Ethiopian Plateau, convergence occurs during Jun to Sep, with the direction of net moisture transport from northeast, as if the net flow over the Plateau originates from the Red Sea and/or the Mediterranean. Detailed analysis of the moisture fields and wind patterns over the basin, at all pressure levels, shows that during Jun to Sep, moisture over the Ethiopian Plateau is largely originated from the Atlantic Ocean, and to a lesser extent from the Indian Ocean, (results are not given here). The moisture is available at all pressure levels and penetrates inland up to the Ethiopian Plateau. On the other hand, the moisture field over the Red Sea is limited only to the lower levels (lower than 850 hPa). The wind patterns over the Ethiopian plateau, up to the 850 hPa level, are from southwest, and it reverses direction at 700 hPa upward (the so-called upper tropospheric tropical easterly jet). Clearly, the topography of the Ethiopian Plateau influences wind direction up to 850 hPa level. A possible interpretation of the Jun to Sep convergence characteristics over the Ethiopian Plateau, is that, moisture over the Plateau (mainly of Atlantic Ocean origin) is lifted up by orography and transported southwest by winds in the upper levels, producing a resultant net moisture transport of a southwestern direction. This doesn't reject that there is moisture transport, at least from the southern part of the Red Sea towards the Ethiopian Plateau. Fig. 6.11 (JJAS) also shows the strong southwesterly monsoon flow over the Somali coast (The Somali jet), a major carrier of atmospheric moisture toward India (Camberlin, 1997). During the winter season Dec to Mar, no convergence occur over the Ethiopian Plateau, in fact considerable divergence takes place.

The White Nile catchment, and in particular the Ghazal basin inhabits a sizeable convergence during Jun to Sep, as well as the area just North of Lake Victoria. The Indian Ocean provides the major source of the summer moisture, in particular to the east of Bahr el Jebel, while the Atlantic moisture contributes to the precipitation over the Ghazal basin. Substantial convergence occurs on the Congo Basin during the two seasons. During the winter season, the White Nile catchment acts as a divergence zone, with some convergence around lake Victoria. It is interesting to note that the convergence areas in Fig. 6.11 correspond to the runoff generating catchments. It is known that there is negligible contribution to the Nile flow from downstream these areas. E.g., the catchment within the Sudan territory has only minor contribution to the Nile runoff. The spatial distribution of the convergence is correlating well with the land topography (see Fig. 3.1).

The regional water cycle over the Nile basin, i.e., both land surface (hydrological) and atmospheric components can be characterized (qualitatively) at the basin level using parameters such as moisture recycling ratio, feedback ratios, precipitation efficiency and moistening efficiency. See Fig. 1.1 for a schematized plot of a regional water cycle. The moisture recycling in a region is the process by which evaporation from the region contributes to precipitation in the same region. The recycling ratio β is defined as the ratio of locally generated precipitation to total precipitation. In the literature different "flavours" of the recycling formula exist. A review of the precipitation recycling formulae and their mutual comparison is given in section 2.3. For the Nile basin we have computed moisture recycling using the formula of Budyko (1974) extended by Brubaker et al. (1993), as given by Eq. (2.1). Savenije (1995), based on a Lagragian approach defined moisture recycling using a moisture feedback ratio ζ, which is the moisture supplied to the atmosphere by evaporation during the wet season relative to precipitation Eq. (2.3). Here the evaporated moisture is not necessarily precipitating back within the same region.

Furthermore, it may be interesting to introduce two more ratios; the precipitation efficiency 'p' and the moistening efficiency 'm'. p is defined as the amount of precipitation in a given region relative to the atmospheric moisture flux overhead, given by:

$$p = \frac{P}{Q} \tag{6.1}$$

Similarly m is defined as the amount of regional evaporation relative to the atmospheric moisture flux (Trenberth, 1999), given by:

$$m = \frac{E}{Q} \tag{6.2}$$

Different definitions for precipitation efficiency also exist in the literature. Schär et al. (1999) defined the denominator of Eq. (6.1) as the total incoming moisture in a region ($E+Q_{in}$) instead of the mean flux Q. It is to be noted that $\zeta=m/p$, when evaporation is negligible outside the wet season, i.e., when $E_w=E$.

It should be stressed that precipitation recycling derived by a regional average formula like Eq. (2.1) serves as a diagnostic measure of the regional land surface-climate interaction. It can be a useful index to compare different basins of the world, but it has no prognostic value. Obviously, the land surface-climate interactions are highly dynamic and nonlinear processes (refer to section 2.3). Only with comprehensive climate modeling simulations it may be possible to obtain a better understanding of these (two way) interactions processes.

The monthly precipitation recycling ratio β, precipitation efficiency p, and moistening efficiency m computed by Eq. (2.1), (6.1) and (6.2), respectively, are shown in Fig. 6.12. The seasonal feedback ratio ζ computed by Eq. (2.3) is given in Table 6.1.

Fig. 6.12 shows that, in accordance with the seasonality of P and E, and since Q_{in} and Q_{out} are fairly steady, both p and m and to some extend β are relatively high during the rainy season. About 40% of the available atmospheric moisture precipitated in the basin during the rainy season, of which around 12% is originated from local evaporation. The local evaporation contributes about 30% of the atmospheric moisture over the Nile during these months. The feedback ratio ζ during the rainy season reaches (E_w/P_w) is 74%, approximately equal m/p. Outside the rainy season, precipitation efficiency drops to about 20%, despite the fact that Q_{in} drops only by about 10% in these months. Obviously a supply of Q_{in} alone is not sufficient to precipitate, a mechanism has to be present (see section 3.1.1).

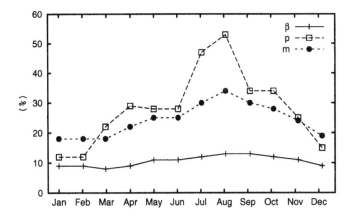

Fig. 6.12: Mean annual cycle of precipitation recycling β, precipitation efficiency p and moistening efficiency m.

It is interesting to compare the annual water cycle over three main river basins: Amazon, Mississippi and the Nile (Fig. 6.13). A similar comparison for the Amazon and the Mississippi is given in Eltahir and Brass (1994). A tabular format of this comparison is given in section 2.4 (Table 2.2). The data on the Amazon are based on ECMWF ERA-15 (Eltahir and Brass, 1994). The data of the Mississippi are based on observations from Benton et al. (1950). The data for the Nile Basin are based on the mean RACMO results 1995 to 2000. The annual fluxes were normalized by the annual precipitation (100%). It is noteworthy that in the literature one may finds different values of precipitation recycling over these basins, depending on: data used, formula applied and size of the basin, see section 2.4 for further details. Using the mean annual data of Fig. 6.13, the precipitation recycling, feedback ratio, precipitation efficiency, moistening efficiency and runoff coefficient are summarized in Table 6.1.

Table 6.1: Parameters of the regional water cycle.

Parameter	Amazon	Mississippi	Nile
Moiture recycling β	17	8	11
Feedback ζ	<58*	n.a	33*
Precipitation efficiency p	83	22	28
Moistening efficiency m	48	17	24
Runoff coefficient C_R	42	22	14

*Upper limit values because ζ is defined as feedback during the rainy season only.

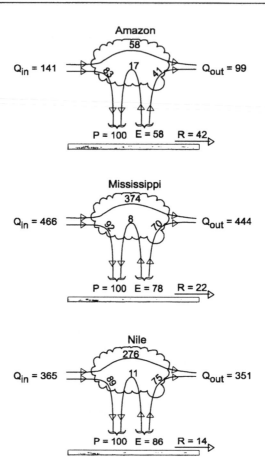

Fig. 6.13: Schematization of the regional annual water cycle over the Amazon, Mississippi and the Nile.

Of the 3 basins, the Amazon shows the largest precipitation recycling within the catchment, followed by the Nile, and the Mississippi. Note that if the formula of Schär et al. (1999) is applied to the same data, β becomes 29%, 14% and 19%, for the Amazon, Mississippi and the Nile, respectively. Qualitatively, this implies that land surface-atmosphere interaction is stronger in the Amazon than in the Nile and Mississippi, respectively. The same is true when considering the moistening efficiency. Although the precipitation efficiency is also decreasing successively in the Amazon, Nile and Mississippi, runoff is not exactly following this sequence. Runoff is the least in the Nile, due to the excessive evaporation in the swamps of the Sudd, Bahr el Ghazal and the Machar marshes. The outflow atmospheric moisture relative to inflow Q_{out}/Q_{in} is 70%, 95% and 96% for the Amazon, Mississippi and Nile, respectively. Although, this ratio is dependent on the large-scale circulation in each basin, the barrier of the Indus Mountain Ranges is likely reducing the outflow moisture.

6.5 Conclusions and discussion

A regional climate model has been applied to the Nile Basin. The model has been customized to simulate the regional climate of the Nile (tropical, semi arid and arid climates). The exercise concentrates on reproducing the regional water cycle as close as possible. Observations on runoff, precipitation, evaporation and radiation have been used to evaluate the model results at the sub-basin level (White Nile, Blue Nile, Atbara and the Main Nile).

The model reproduces runoff reasonably well over the Blue Nile and Atbara sub-basins, while overestimates the White Nile runoff. The extremely small runoff coefficient and huge catchment area of the White Nile makes the runoff very sensitive to inaccuracy of precipitation or evaporation. Except for the period Mar to Jun over the White Nile, the model simulates the precipitation well over the 4 sub-basins, in particular the time variation. The underestimation of precipitation on the White Nile during Mar to Jun is partly related to the ERA-40 forcing on the southern boundary of the model. The evaporation over the Sudd wetland could be accurately simulated during the rainy season, while it was overestimated during the dry months because permanent flooding is assumed. In fact, the largest part of the Sudd is a seasonal swamp. The soil moisture result over the Sudd is compatible with evaporation results, i.e., closely resembles remote sensing derived estimates during the wet period, and underestimated during the dry months. Limited observations on radiation (2 stations) were compared to model results. The model overestimates the incoming short wave radiation for some months, while producing compatible results of the incoming long wave radiation.

Subsequently, the model has been used to compute the regional water cycle over the Nile Basin. The mean annual moisture recycling over the basin has been computed by the Budyko formula as 11%; monthly values vary between 9 to 14%. The annual results on the Nile water cycle have been compared to the Amazon and the Mississippi data given in the literature. The moisture recycling is 17, 8 and 11% over the Amazon, Mississippi and the Nile, respectively, while the precipitation efficiency is 83, 22 and 28%, respectively. The annual runoff coefficient over the 3 basins is 0.42, 0.22, and 0.14, respectively. This clearly shows that land atmosphere interaction over the Nile (and Mississippi) is much less pronounced as compared to the Amazon. Although the comparison between the 3 basins show interesting conclusions on the relative water cycle components among the basins, the bulk recycling ratio alone is not sufficient to provide an in-depth understanding of the land surface-climate interaction processes. These processes are dynamic, and highly nonlinear in nature. It is inaccurate to derive conclusions on the impact of regional precipitation based on, e.g., alteration of β. A more realistic approach to study the impact of land use changes on regional climate would be through applications of a sound RCM. In this particular case, the Nile model is used to simulate a second scenario of a dried Sudd wetland, and subsequently studying the impact on regional hydroclimatology as given in Chapter 7.

A regional atmospheric model calibrated against flow regimes and distributed remote sensing data is a strategic tool for understanding the impacts of climate change on water management and vice versa. In view of the growing problem of water scarcity, the demand for advanced atmospheric-hydrological tools – such as RACMO – is growing.

7. The Impact of the Sudd wetland on the Nile Hydroclimatology[1]

7.1 Introduction

In recent years, climatology and hydrology are increasingly merged into one discipline – in particular at a river basin level – due to the coupled nature of the land surface-atmosphere interactions. It is widely believed – through observations and modeling studies – that continental land use changes can impose changes on regional climate, and simultaneously climate change has direct implications on basin hydrology and water resources. In the international context, a tantamount of research and experiments have been carried out and still undergoing to define the land surface-atmosphere interactions, and the possible impact of land use change on climate (see Chapter 2, and the review by Garratt, 1993; Betts et al., 1996; Viterbo, 2002). However, the theme is still challenging and awaits a much deeper comprehension, as no consensus is yet reached among different researchers on the possible feedback mechanisms (Chapter 2; Koster et al., 2002).

Many studies support a positive soil moisture-atmosphere feedback, which may occur at varying time scales, from diurnal to seasonal. A positive feedback implies that an increased soil moisture anomaly favors an increase of precipitation. Eltahir (1998), Schär et al. (1999), and others, hypothesized that soil moisture enhances energy supply (latent and sensible heat) to the atmosphere, and accompanied by a reduced depth of the planetary boundary layer, results into an increase of the moist static energy. This effect, if occurring over a large enough region, would favor both local convective storms and large-scale circulations. Using one dimensional climate model Ridder (1999) showed that the potential for moist convection increases with the evaporative fraction (i.e., with soil wetness). For the Mississippi basin, Betts et al. (1999) showed that wet soil produces larger equivalent potential energy, which probably produces a positive feedback, although it may not control the flood precipitation over the Mississippi. Over Europe, Schär et al. (1999) identified positive soil moisture-precipitation feedback using RCM simulations. In West Africa, Zheng and Eltahir (1998) showed a positive soil moisture rainfall feedback when imposing a large-scale soil moisture anomaly in their numerical experiments. A decrease of local rainfall in the wetlands of South Florida due to land use change was shown by Pielke et al. (1999) using RCM experiments. The land use change mainly concerns the type and amount of vegetation cover. Similarly, using basin average moisture recycling formulae, different researchers indicated a positive soil moisture-atmosphere feedback, e.g.; Brubaker et al. (1993), Savenije (1995), Trenberth (1999).

---

[1] Based on: Mohamed, Y.A., van den Hurk, B.J.J.M., Savenije, H.H.G., Bastiaanssen, W.G.M. 2004d. The Impact of the Sudd wetland on the Nile Hydroclimatology, accepted in the J. of WRR.

Other researchers who claim a negative soil moisture feedback, i.e., decreased soil moisture anomaly favors a reduced rainfall. Over the Mississippi, Giorgi et al. (1996) showed that increased soil moisture causes a decrease of rainfall, explained by a reduced evaporation (dry soil condition) increases the buoyancy, which dynamically sustains convective precipitation. Pan et al. (1999) showed a decrease in rainfall for an increased evaporation in the central US, attributed to weakening of dynamic forcing required for precipitation. Using a one-dimensional climate model, and under specific atmospheric conditions, Ek and Holtslag (2004) showed that (in some cases) a decreasing soil moisture may actually lead to an increase in clouds. Findell and Eltahir (2003) derived both positive and negative soil moisture-atmosphere feedback over the US, depending on the region and the oceanic influences.

It is clear that it is inappropriate to extrapolate the derived feedback conclusions in space and time, in particular for land use climate impact studies (Ek and Mahrt 1994; Pan et al., 1999). Different models may give different feedback results (even different in sign) on the same basin, as demonstrated for instance for the evaluation of the impact of the Amazon deforestation on the basin hydroclimatology reviewed by Lean and Rowntree (1997). Further complications arise when considering the magnitude of the land use change impact on climate. Often it is difficult to separate a potentially small change signal from a potentially large level of noise. One final comment on land use change climate studies is that, most of the numerical experiments imposed exaggerated changes in order to obtain statistical significance of the results. E.g., the soil moisture anomaly in West Africa by Zheng and Eltahir (1998) was imposed over an exceptionally large area. Noticeably, different results were obtained in the same region when realistic land use change was imposed. Taylor et al. (2002) indicated that the land use change is not the principal cause of the recent drought over the Sahel (West Africa), which was claimed by many climate modelling studies. They have estimated the land use change between 1961 and 1996 to be about 4% (conversion of the Sahel land from tree cover to bare soil). This caused a reduction of rainfall by 5% based on GCM simulations. They argued that this land use change is not large enough to have been the principal cause of the recent drought in the Sahel.

Similarly, the literature shows no consensus for the impact of the Sudd wetland on the regional climate. It may well be that draining the Sudd swamps will reduce the precipitation within the Nile basin, thereby reducing the net impact on the Nile flows. The Jonglei investigation Team (JIT, 1954) suggested that drying part of the Sudd wetlands would cause no reduction of rainfall, since reduction of moisture supply to the atmosphere will be compensated by an increased convection caused by the dried swamps. Howell et al. (1988) and Sutcliffe and Parks (1999) argued that no decrease of the regional rainfall is expected after completing Jonglei canal (partial draining of the Sudd), because conversely there was no rainfall increase due to the expansion of the Sudd swamps (double in size) following the increased Nile flow in the early 1960's. Fig. 7.1 shows the sudden increase of the Sudd inflow around 1961/64, which was correlated with an increase of the swamp area due to the mild slope of the terrain. However, as depicted by the bottom panel (o), the rainfall didn't show any associated increase in that period.

Eltahir (1989) hypothesized that a decrease of the Bahr el Ghazal swamp area (smaller than the Sudd) will reduce rainfall in the central Sudan. However, the hypothesis could not be proved by the available data. Kite (1998) citing the work of Eagleson (1986), who claimed that evaporated water from the Sudd falls as rain in other areas of Africa, with some water reaching Europe and South America. It can be concluded that despite the importance of the Sudd wetlands, both to the local environment and as an expected additional supplier of the Nile water resources – directly or through moisture recycling – there is no clear and unified answer to the question: what is the effect of the Sudd wetlands on the regional hydrological cycle terms, and hence on the amount of water transported through the Nile?

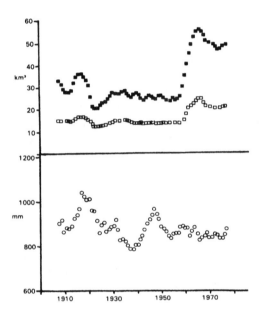

Fig. 7.1: Top panel Sudd inflow (■) and outflow (□) in Gm³/yr, compared with (bottom panel) rainfall over the Sudd (○) in mm/yr. The data represent 7 year moving average (from Howell et al., 1988).

In this chapter we have applied a RACMO model (described in Chapter 6) to study the impact of the Sudd wetland on the Nile hydroclimatology. The objective is to quantify the impact of the drained Sudd both on local climate and on the regional water cycle. RACMO has been adjusted to represent the present regional climatology (Control run CTL given in Chapter 6). The same model is also run for a drained Sudd scenario (Drained run DRA). The differences between the two scenarios are quantified to evaluate the impact of the Sudd on local and regional climate. Description of the Nile hydroclimatology and the Sudd wetland characteristics is given in Chapter 3.

7.2 The Sudd experiment

The experiment is basically designed to evaluate the impact of the Sudd wetland on the Nile hydroclimatology. First, a RCM based on RACMO has been calibrated to simulate the present climatology including the Sudd wetland (CTL run) reported in Chapter 6. Subsequently the same model is applied to a modified land use scenario, i.e., a drained Sudd scenario (DRA run).

7.2.1.   The control run

The model has been adjusted to simulate the spilling of the Nile over the Sudd by routing the flow from the upstream catchment to the 15 grid points of the wetland. The spill R_{in} is supplied everyday directly to the soil moisture. By adopting a very low minimum canopy resistance (15 s/m), it was possible to mimic the evaporation of Nile water inundation over the wetland. The model has been validated against observational data sets of radiation R_n, precipitation P, runoff R, and evaporation E. A detailed evaluation of the model results at the sub-basin level (Blue Nile, White Nile, Atbara River and the Sudd) is given in section 6.4. In general, the model could capture the wet season Jun to Sep condition over the sub-basins reasonably well. Runoff from the Ethiopian Plateau is fairly reproduced by the model, while it overestimates the White Nile runoff at Malakal and further downstream. Over the Sudd, P, E, and dS/dt (change of soil moisture content) were accurately reproduced during the wet season, while E was overestimated during the dry season Dec to Mar. It has been concluded that the model provides a sound representation of the hydroclimatological processes over the region, in particular during the wet season.

7.2.2.   The drained Run

The second scenario (DRA) is run with exactly the same model configuration and boundary condition, with the exception that the Nile does not spill over the Sudd. Thus, the primary land use change over the Sudd would be less available moisture for evaporation in DRA compared to CTL run.

Fig. 7.2 depicts results from the two runs of the mean hydrological cycle averaged over the Sudd. It shows that E is high in the dry season for the CTL run, while it is virtually zero for DRA. The difference ΔE reaches more than 5 mm/day in the dry season, while it is small in the wet season. Rainfall is a major source of wetland flooding during the wet season, thus E_{DRA} approximates E_{CTL}. Unlike DRA, the steady and uniform supply of R_{in} provides sufficient soil moisture for evaporation in the CTL run, in particular during the dry season. This enhances the seasonality of dS/dt in the CTL run. While, for DRA the input and output follows the same seasonality, which results in a mild seasonality of dS/dt. The runoff in the DRA run is almost zero, while some runoff is generated in the CTL run, although smaller than the observed amount of about 1.5 mm/day.

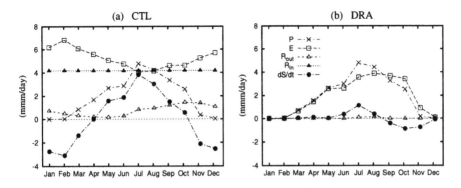

Fig 7.2: The simulated hydrological budget terms averaged over the Sudd wetland, (a) Control run, (b) Drained run. (Mean annual cycle 1995 to 2000)

7.3 Further analysis of the simulations

The soil moisture and evaporation difference between the two scenarios displayed in Fig. 7.2 have several implications at varying temporal and spatial scales. In this section we discuss the effects on local climate (temperature and relative humidity), and the modification caused to the regional water cycle. Subsequently, the change of the Nile flow caused by draining the Sudd is estimated.

7.3.1. Impact on micro climate

The difference of the relative humidity at screen level (2m above the ground) between the two runs ΔRH is shown in Fig. 7.3 for the two seasons: the dry season (Dec to Mar) and the wet season (Jun to Sep), respectively. The significance of ΔRH is evaluated by normalization with one standard deviation of the six seasonal values from the 1995-2000 CTL simulation ($\sigma_{RH(CTL)}$). The ratio $\Delta RH/\sigma_{RH}$ is a measure of the statistical significance of the mean ΔRH against the natural variability. It is assumed that $\Delta RH/\sigma_{RH} \geq 1$ indicates a significant change. Assuming normal distribution, for 2/3 of the data the change caused in RH is higher than the natural variability quantified by σ_{RH} within the 6-year record. This statistical measure is adopted for the other climate parameters analyzed in this study. Over the Sudd ΔRH reaches 30 to 40 % in the dry season, while it is small (<10 %) in the wet season. For both seasons, the change over the Sudd is significant (large $\Delta RH/\sigma_{RH}$). Except for somewhat small patches on the eastern part of the Congo Basin in the dry season, and west of the Marra Mountain (24°E, 15°N) in the wet season, ΔRH is negligible outside the Sudd. The spatial plots of ΔRH at higher elevations show negligible values at levels higher than 850 hPa (approximately 1,500 m from the ground). This indicates that the supply of evaporative moisture from the Sudd is limited to lower levels.

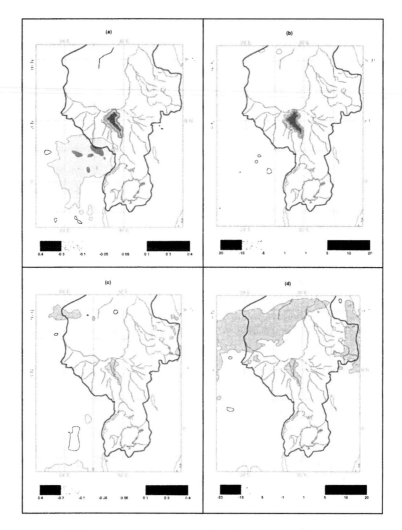

Fig.7.3: (a) Contours are dry season values of mean ΔRH, (b) dry season values of $\Delta RH/\sigma_{RH}$, (c) and (d) as (a) and (b) for wet season.

The reduced evaporation (latent heat) for the DRA run enhances the sensible heat over the Sudd. This raises the screen level temperature by 4 to 6°C in the dry season in the middle of the Sudd, reduced to about 0.5 °C at the boundary (Fig. 7.4a). There is a small reduction in the temperature on the eastern Congo Basin, but insignificant with regard to the ratio $\Delta T/\sigma_T$. In the wet season, ΔT is small, around 0.5 to 1°C, and insignificant. While the evaporation difference is limited only within the wetland territory, the spatial distribution of ΔT expands to a somewhat larger area, mainly to the west and southwest of the Sudd, influenced by the westward wind pattern in the dry season. The seasonal range of the temperature over the Sudd in the CTL run is small (2 to 3 °C), while this range increases to about 4 to 6 °C for the DRA run.

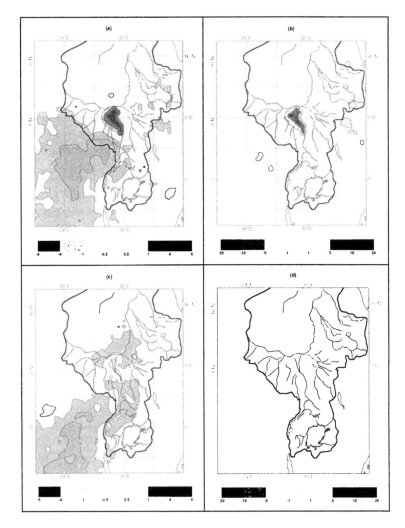

Fig.7.4: (a) Contours are dry season values of mean ΔT, (b) dry season values of $\Delta T/\sigma_T$, (c) and (d) as (a) and (b) for wet season.

7.3.2.   Impact on regional climate

The convergence/divergence of the atmospheric moisture flux is a key term in the regional water cycle. Refer to section 6.4.5 for a description of the atmospheric moisture fluxes, their sources and main directions of movement in the Nile region. The balance of the atmospheric part of the water cycle, temporally and spatially averaged over a region:

$$\left\{\overline{\nabla Q}\right\} = \left\{\overline{P}\right\} - \left\{\overline{E}\right\} + \left\{\overline{\frac{dW}{dt}}\right\} \qquad\qquad (7.1)$$

where ∇Q is the convergence/divergence in a model grid. Q is computed as $Q = \int q\vec{u}dz$, i.e., the vertical integral of $q\vec{u}$, where q is the specific humidity and $\vec{u}$ is the wind speed vector. dW/dt is the change of atmospheric moisture content, and it is very small compared to Q, –2% in the dry season and 1% in the wet season. However, dW/dt is relatively larger during the transition period from dry to wet +3%, and –3% from wet to dry. On annual basis dW/dt approaches zero, and hence $\{\nabla Q\}$ equalizes $\{P\}$-$\{E\}$, which in turns equals the runoff from the basin, presuming negligible ground water storage change when averaged over a couple of years (Seneviratne et al., 2004).

Fig. 7.5a and 7.5b show the horizontal atmospheric moisture fluxes Q in magnitude and direction represented by the arrows, overlaid on the convergence/divergence field ∇Q. The data are mean seasonal values (1995-2000) of the CTL run sampled from 12 hourly output of q and $\vec{u}$. The moisture transport in the dry season is from east to west over the White Nile and Sudd, indicating moisture of Indian Ocean origin. Over the Ethiopian Plateau (sources of the Blue Nile and Atbara River), there is negligible atmospheric moisture transport during the dry season. In this season divergence occurs over the Ethiopian Plateau and from major parts of the White Nile including the Sudd. The strong divergence over the Sudd is due to excessive evaporation from the wetland (see Fig. 7.2a). There is considerable convergence in the eastern part of the Congo Basin (22°E, 0°N) in the dry season.

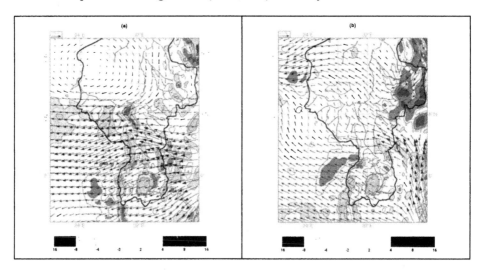

Fig.7.5: The atmospheric moisture flux (kg/m/s), and convergence/divergence filed (mm/day) for the CTL run, mean of 1995 to 2000: (a) dry season Dec to Mar, (b) wet season Jun to sep.

During the wet season (Fig. 7.5b), moisture transport over the White Nile and Sudd is from east and southeast, while over the Ethiopian Plateau the transport is from northeast. It is worth mentioning that the direction of net moisture transport varies significantly with height. There is large convergence over the Ethiopian Plateau, which results in the high seasonal precipitation and runoff of the Blue Nile and the Atbara River. Over 70% of the Blue Nile flow is generated during the wet season. Convergence also occurs north of lake Victoria, over the Baher el Ghazal sub-basin, and near the Nuba hills (28°E, 11°N). Over the Sudd area precipitation and evaporation are nearly balanced (see Fig. 7.2a), and convergence is consequently small.

The impact of the Sudd wetland on atmospheric moisture fluxes and convergence/ divergence can be seen from difference plots of Q and ∇Q. The significance of the mean difference is evaluated by normalization with one standard deviation of the seasonal values from the CTL simulation, $\sigma_{\nabla Q(CTL)}$. However, small numerical errors in the calculation of ∇Q from the 12-hourly $\vec{u}$ and q fields become relatively important when calculating difference plots of ∇Q. Therefore, $(\nabla Q_{DRA} - \nabla Q_{CTL})$ was calculated as $(P-E+dW/dt)_{DRA} - (P-E+dW/dt)_{CTL}$, see Eq. (7.1).

Fig. 7.6a and 7.6b show the results for the dry season, and Fig. 7.6c and 7.6d for the wet season, respectively. Significant alteration of ∇Q occurs over the Sudd in the dry season (Fig. 7.6a and 7.6b). Note that the scale of atmospheric fluxes (arrow) is 10 times bigger than the scale given in Fig. 7.5. The ∇Q difference over the Sudd reaches more than 5 mm/day in the dry season, due to excessive evaporation from the inundated wetland in the CTL run. Except for few patches in the eastern Congo Basin, there is hardly any significant difference of the convergence/divergence fields in the rest of the model domain in the dry season.

During the wet season, evaporation from the Sudd does not affect the regional hydrology as much, since almost all the surrounding areas are already wet from the rain (see ΔRH in Fig. 7.3c and 7.3d). Except for the small patch at the middle of the Sudd, the plots of Fig. 7.6c and 7.6d show hardly any significant change, neither locally nor in the Nile Basin. Small negative patches occur on the eastern Congo Basin, resulting from changing net convergence in the CTL run (due to individual convection events) to net divergence in the DRA run.

Though not significant, change of the horizontal flux (Q) is opposite to the direction of the mean transport in the CTL run over the Sudd for both seasons because of evaporation. Also insignificant, ΔQ indicates a south north gradient in areas west of the Nile. A possible explanation is that air passing over the drained Sudd has higher water uptake capacity, which seems to enhance water advection from further south, where enhance surface evaporation provides the source of this water flux. This is compatible with earlier discussion on ΔRH and ΔT given in section 7.3.1 (see Fig. 7.3 and 7.4), i.e. the enhanced evaporation increases RH and reduces T. In the dry season, the area on the eastern Congo shows a small, though not significant increase of RH and a slight decrease of T.

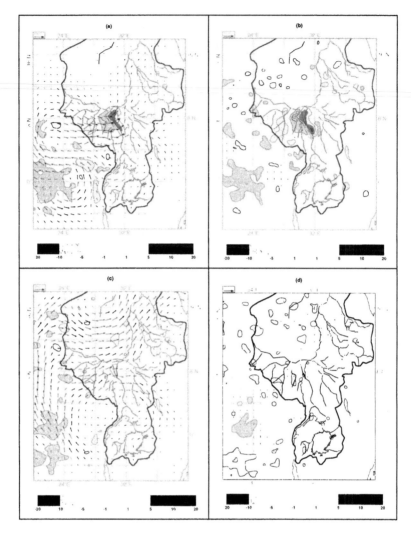

Fig.7.6: (a) Contours are dry season values of mean $(\nabla Q_{DRA}-\nabla Q_{CTL})$, arrows are $(Q_{DRA}-Q_{CTL})$, (b) dry season values of $(\nabla Q_{DRA}-\nabla Q_{CTL})/(\sigma_{\nabla Q(CTL)})$, (c) and (d) as (a) and (b) for wet season.

7.3.3.   Implication on the Nile hydrological cycle

To distinguish the final impact on the Nile Basin budget, we present the mean annual time series of P, E, R, and dS/dt of the Nile catchment at the Aswan outlet. Fig. 7.7 shows the time series of the two runs, together with one standard deviation of the inter-annual variability. The difference in P, E, R and dS/dt, is much smaller than σ. Detailed inspection of the model results at smaller time steps (6 hourly) shows that changing conditions in the Sudd area triggers erratic intense rainfall events far away from the Sudd (e.g., in the Atbara catchment). However, when average over 6 years period, the mean annual cycle shows a systematic (slight) increase in P and E. This suggests that enhanced convection over the dry soils of the DRA run causes the systematic increase in P. This mechanism was also suggested by the JIT (1954) for the Jonglei canal (see section 7.1). The runoff from the two runs is almost similar, indicating that the increased precipitation is mainly consumed by evaporation, after altering the soil moisture content. It is to be note that on an annual scale the soil moisture in the basin is not a sink or source.

Similar time series comparisons have been made for the sub-basins (Blue Nile, White Nile, Atbara River). However, results have similar characteristics as given in Fig. 7.7 for the Nile.

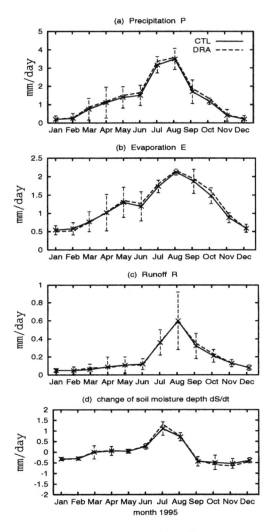

Fig. 7.7: The hydrological budget terms of the Nile at Aswan (mm/day), mean annual cycle 1995-2000: The error bars represent one σ of monthly mean values during the 6 simulation years.

7.3.4.   Impact on moisture recycling

Further confirmation to the results obtained on the impact of the Sudd on the regional hydrological budget can be approached through the analysis of the mass balance terms and moisture recycling on a smaller region centered around the Sudd (2°N to 12°N, 26°E to 36°E; see Fig. 3.3). This region is considered large enough to allow detection of the impact of evaporation changes in the Sudd on moisture recycling, and small enough to distinguish the evaporated water volume from the water volume transported by the atmosphere. The results of the hydrologic budget time series P, E, R and dS/dt, are qualitatively similar to the results of the whole Nile given in Fig. 7.7 (not shown here).

The effect of local evaporation on the regional hydrological cycle can be diagnosed from a number of dimensionless indices: the precipitation efficiency (p), moistening efficiency (m), recycling ratio (β), and feedback ratio (ζ). The precipitation efficiency quantifies how much of the atmospheric moisture crossing the region is precipitated Eq. (6.1). The moistening efficiency describes how much the local evaporation contributes to the atmospheric moisture flux Eq. (6.2). The recycling ratio quantifies how much of the evaporated moisture is precipitated back into the same region Eq. (2.1). The feedback ratio quantifies the relative magnitude of the wet season evaporation E_w to precipitation, which is not necessarily precipitating back in the same region Eq. (2.3). See section 2.3 and 6.4.5 for further details on these parameters.

The mean monthly variation of the three terms p, m and β for the two runs is shown in Fig. 7.8. A slight increase of p, m, and β is shown for the DRA run, although small compared to the inter-annual variability. During the peak rain Jul to Sep, when the whole area around the Sudd is wet, m and β are almost similar for the two runs. During the wet season the mean feedback ratio ζ is 82% and 79% for the CTL and DRA runs, respectively, owing to a higher precipitation rate in DRA. The relatively small impact of draining the Sudd on regional hydrological cycle can be explained partly by the low values of β and ζ in the dry season.

In general, these results are compatible with the times series and map results given in the previous sections. Draining the Sudd has a minor impact on the regional water cycle, and it shows a slight increase of the budget terms, though insignificant compared to the inter-annual variability.

Although the evaporation rate over the Sudd is about 3 times the average rate in this region, volume wise the Sudd evaporation is only 7% of the region evaporation. Furthermore, the Sudd evaporation constitutes only 1% of the volume of the atmospheric moisture transport in the region. Obviously these results are expected considering the small area size of the Sudd relative to the region (about 2.6%). So, it can be concluded that in terms of mass balance, the impact of the Sudd on the water budget in the Nile Basin is negligible.

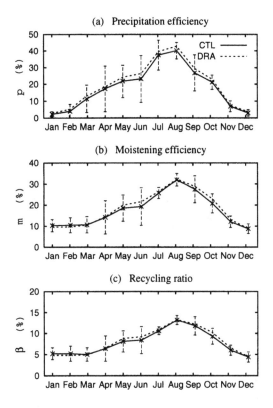

Fig. 7.8. (a) Precipitation efficiency *p*, (b) Moistening efficiency *m*, (c) Recycling ratio *β*, for the two runs over the region (2°N, 12°N and 26°E, 36°E), (mean annual cycle 1995-2000). The error bars represent one σ of monthly mean values during the 6 simulation years.

7.3.5.   Implication on the Nile water resources

The previous results indicated that the impact of the Sudd wetlands on the regional hydrological budget is small and insignificant relative to the inter-annual variability, while large changes occur to microclimate in our DRA experiments. The additional water provided at the outlet downstream of the Sudd is the whole river runoff generated upstream of the Sudd for the DRA run. While for the CTL run, the major part of this runoff is evaporated over the Sudd, in particular during the dry season. We evaluate the effect of this spill on the discharge at Malakal, located just downstream the Sudd (catchment area is $1.48 \ 10^6 \ km^2$).

As mentioned before, instead of the direct RACMO output, the observed Nile flow upstream of the Sudd has been used as spill over the 15 grid points of the wetland. This was done because RACMO clearly underestimates the runoff of the basin upstream. The amount of water spilled over the Sudd is 53 Gm^3/yr (49 Gm^3/yr being the 1961-1983 long term mean), whereas RACMO produced only 21 Gm^3/yr runoff from the upstream catchment area. With the additional 32 Gm^3/yr, the CTL run runoff simulated at Malakal is 64 Gm^3/yr. In the DRA run, the runoff generated at Malakal is 78 Gm^3/yr. This amount should be corrected in a similar way as in the CTL run to allow comparison. In the DRA run, the runoff generated in the upstream area is 24 Gm^3/yr (+3 Gm^3/yr). Assuming that a similar correction for underestimating the upstream runoff should be applied to the DRA run, the runoff downstream of the Sudd at Malakal amounts to 78+32=110 Gm^3/yr, 46 Gm^3/yr more than in the CTL run. However, as shown in Fig. 6.8 the simulated evaporation in the CTL run is higher than observed in the dry season. We estimate that this overestimation amounts to 10 Gm^3/yr. So, a correction of the CTL runoff by 10 Gm^3/yr seems reasonable. In the DRA run there is no overestimation of evaporation in the dry season, because of no river spill (see Fig. 7.2b). This implies that the "best guess" on the effect of draining the whole Sudd wetlands leads to a net gain of 46-10=36 Gm^3/yr, about half the Nile flow at Aswan (see Fig. 7.9). This is somewhat more than the long-term 1961-1983 mean losses over the Sudd of 29 Gm^3/yr computed by Sutcliffe and Parks (1999). This discrepancy is well understood from the noted model deficiencies and bias corrections applied.

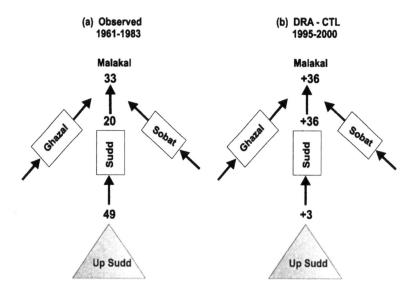

Fig. 7.9: Gain of the Nile water at Malakal in Gm^3/yr

7.4 Conclusion and discussions

Evaluation of the results from the two simulations (with and without the Sudd) has shown that draining the Sudd can have large impact on microclimate over the Sudd. The reduced evaporation in the drained scenario causes a reduction of the screen level relative humidity by 30 to 40 %, and an increase of the temperature by 4 to 6 °C during the dry season. The microclimate impact is confined to the Sudd area and the adjacent area to the west and southwest direction influenced by the wind pattern in the dry season. During the wet season the impact of the Sudd drainage is small, because the surrounding area is already saturated by the rain. However, a slight decrease of relative humidity (<10%) is detected over the Sudd in the wet season.

The simulated results show that the impact of the Sudd draining on the regional water cycle (atmospheric moisture fluxes, precipitation, evaporation, runoff and sub-surface storage) is negligible and insignificant compared to the inter-annual variability of these parameters. In terms of mass balance, the Sudd evaporated volume is a small fraction relative to the oceanic moisture in the region (less than 1% of the atmospheric moisture in the Nile Basin). However, the drained Sudd may create some climatic disturbances, which in our simulations generate slightly higher rainfall than the present situation. This is most likely attributed to an enhanced convective activity that occurred over the dried wetlands. During the wet season, influence of the Sudd is not noticed. In the dry season, when the Sudd may cause the largest impact on the water cycle, there is no mechanism for rainfall, at least in major parts of the basin (the ITCZ is retreated, and the upper troposheric tropical easterly jet disappeared).

The net gain of the Nile water by complete draining of the Sudd wetland would be an additional ~ 36 Gm^3/yr (relatively higher than observed Nile losses over the Sudd of 29 Gm^3/yr). This is slightly less than half the long-term mean of the Nile natural flow at Aswan. The famous Jonglei canal phase I would save about 4 Gm^3/yr (5% of the Nile flow at Aswan), and drain about 30% of the Sudd wetland. Most likely there will be no significant impact on the regional hydrological budget terms, while the impact on the microclimate over the dried parts would be comparable to the derived results in this study.

The strength of the Sudd experiment presented in this chapter is that it utilizes a coupled land surface atmosphere model incorporating all possible feedback mechanisms to address a water resources question. The weakness of the experiment is that its time span is limited to only 6 years, making it hard to detect the actual signal caused by land use change from the natural variability in the record. Obviously, one would expect improved results from a longer simulation (e.g., 40 yr) and a finer resolution model (smaller than 50 km grid). However, the main conclusions derived here are likely to be confirmed rather than discarded, since the hydrological fluxes to and from the Sudd are relatively small compared to the atmospheric fluxes.

Anxiety on the impact of the Jonglei canal has been high with regard to reduction of regional rainfall due to reduced evaporation from the Sudd. This study does not confirm this. However, the impact on microclimate and the regional distribution of water resources will be quite significant.

8. Summary and Conclusions[1]

The noble civilizations developed along the Nile (Pharaonic, Nubians, etc.) were flourishing because (mainly) of the achievement they have had in irrigation works that sustained reliable food supply. At that time, agricultural production depends on how high the flood would be. Off the flood season, the Nile water passes down to the Mediterranean Sea. Mounting population pressure, urges the needs for utilizing all seasons flow, which lead to the building of storage reservoirs during the 19th and 20th centuries. With further increasing demands, planners have had the intention to increase the all seasons flow by reducing evaporation from the Upper Nile swamps (Sudd, Bahr el Ghazal, Sobat Marches), and pass water directly into the river further downstream.

Is the evaporated moisture from these Upper Nile swamps, contributing to rainfall in the Nile Basin? A question, although raised simultaneously with the plans for draining these swamps – as early as the beginning of the 20th century – the answer was always obscure, and never precise. The supportive parties of draining the swamps, guessed no negative impact on climate, the anti-parties expressed high anxiety of a reduced rainfall, also based on hypothesis.

To answer this question, first the actual evaporation E_a from the entire Sudd wetland ecosystem needs to be precisely derived. The literature shows a wide range (1533 to 2150 mm/yr, or even higher). The impact of evaporation on climate, could it be quantified by a basin average moisture recycling formulae, i.e., by addressing the mass balance of the water cycle alone – which save a lot of computation effort – or should the whole land surface-atmosphere feedback loops be accounted for? These are the research questions addressed in the thesis. The objective is to better understand the Nile hydroclimatology, to quantify the impact of the Sudd wetlands both on the local and regional climate conditions, and hence assess the implications of wetland management on the Nile river discharges.

The Sudd wetland is a huge swamp on the upper Nile with very scanty data (physically and politically inaccessible), in which case remote sensing becomes an instrumental technique to estimate E_a from these vast wetlands. More than 115 NOAA-AVHRR images (1 km resolution) covering 3 years – of different hydrometeorological conditions – have been processed with the SEBAL algorithm. SEBAL is a Surface Energy Balance Algorithm for Land that converts spectral radiances into energy balance fluxes during specific cloud-free days (section 4.2.2). The objective was to estimate the monthly actual evaporation over the upper Nile wetlands. The image covers an area of 1000 km x 1000 km, including the swamps of the Sudd, Baher el Ghazal and the Sobat sub-basins. The areal average of E_a value for the Sudd, Ghazal, and Sobat are 1636, 1499 and 1287 mm/yr, respectively (section 4.3). E_a for Ghazal and Sobat were averaged over catchments downstream the gauging stations. The Sudd wetland area A was delineated based on evaporation

---

[1] Based on: Mohamed, Y.A., Savenije, H.H.G., Bastiaanssen, W.G.M., van den Hurk, B.J.J.M., New lessons on the Sudd hydrology learned from remote sensing and climate modeling, to be submitted to HESS.

characteristics for a medium year (2000) and amount to 38,000 km^2 (Fig. 4.6). The two figures of E_a and A derived for the Sudd are largely different from the most recent estimates, e.g., in the Jonglei canal studies (Table 4.6).

Reliability of the derived E_a estimates has been verified through two approaches: water balance calculations over the wetlands, and analysis of the biophysical properties that control evaporation. The water balance calculations show close resemblance of E_a for two of the three sub-basins. The closure terms in the annual water balance were 1.8% and 5.7 % for the Sudd and the Sobat sub-basins (Table 4.3 and 4.5, respectively). The balance does not close for the Bahr el Ghazal sub-basin, due to considerably small inflow, attributed to un-gauged catchment runoff (Table 4.4).

The biophysical properties of the swamp vegetation derived from satellite measurements: albedo, leaf area index, surface resistances, roughness height, and evaporative fraction, are compared to the literature values. The values are found to lay within the expected ranges (Fig. 5.7). The comparison also includes the ratio of E_a/E_w (wetland evaporation to open water evaporation), in which case it is demonstrated that the concept of E_a proportional to E_w does not hold true for wetlands (Fig. 5.1). The temporal variability of the biophysical properties over the Sudd showed that the variation of the atmospheric demand in combination with the inter-annual fluctuation of the groundwater table – due to a pronounced seasonality in rainfall and inundation by the Nile floods – results into a quasi-constant evaporation rate in the Sudd throughout the year (section 5.3). A crude calculation of the groundwater table – based on satellite measurements and water balance calculations – show that the Sudd wetland has a profound seasonal variation of the groundwater table, which demonstrates that a majority of the Sudd is non-inundated and has a seasonal decaying vegetation system (Fig. 5.10). It has been shown that the Penman-Monteith equation can provide a solid physical basis, but the spatial distributed biophysical properties must be known, which could be provided from remote sensing techniques.

Could a regional average moisture recycling formulae quantify the impact on climate caused by a land use change (e.g., draining the Sudd) ? An intensive literature review, including an international electronic discussion forum was performed to review the state-of-the-art of the research on moisture recycling. This is presented in Chapter 2. Results in the literature from selected case studies are presented and discussed (section 2.4). The commonly used methods on moisture recycling were applied to the annual fluxes of four large basins: Amazon, Mississippi, Nile and the West Africa region (Table 2.2). It is found that large variation between various methods and results exists. Although these methods can be useful for qualitative comparison between the basins, they are incapable of quantifying the processes of land surface-atmosphere interactions at varied temporal and spatial scales. These processes are dynamic, and highly nonlinear in nature. The water balance approach used in these formulae is too simple to define accurately the physics of the land surface–climate interaction. Therefore, it is a must – despite the computational effort – to use a coupled land surface regional scale atmospheric climate model to quantify the impact of land use change on climatic state variables.

A regional climate model has been applied to the Nile Basin for two scenarios: with Sudd (CTL run), and a drained Sudd (DRA run). The difference between the two scenarios has been quantified to evaluate the impact of the Sudd on the Nile hydroclimatology.

The regional climate model is based on RACMO, and located between 10°E to 54.4°E and 12°S to 36°N, has 0.5° horizontal resolution, and 31 vertical levels. The initial and lateral boundary conditions are based on ERA-40 reanalysis data. The simulation period extends from 1995 to 2000. An especial adjustment has been made to the RACMO model to incorporate routing of the Nile flow that inundates the Sudd. This is incorporated simultaneously (every day) to simulate the actual flooding over the Sudd, and by fixing a low surface resistance it was possible to mimic evaporation characteristics from wetlands.

Observations on runoff, precipitation, evaporation and radiation have been used to evaluate the model results at the sub-basin level (White Nile, Blue Nile, Atbara and the Main Nile). The model reproduces runoff reasonably well over the Blue Nile and Atbara sub-basins, while overestimates the White Nile runoff (Fig. 6.4). The extremely small runoff coefficient and huge catchment area of the White Nile makes the runoff very sensitive to inaccuracy of precipitation or evaporation. The model simulates the precipitation well over the 4 sub-basins, in particular the time variation, except for the period Mar to Jun over the White Nile (Fig. 6.5). The evaporation over the Sudd wetland could be accurately simulated during the rainy season, while it was overestimated during the dry months because permanent flooding is assumed (Fig. 6.8). Limited observations on radiation (2 stations) were compared to model results. The model overestimates the incoming short wave radiation for some months, while produces compatible results of the incoming long wave radiation (Fig. 6.9). It has been concluded that the model provides a sound representation of the hydroclimatological processes over the region.

Subsequently the model is run for the drained Sudd scenario (DRA run). It is exactly the same model configuration and boundary condition, except the Sudd is being completely drained. Thus, the primary land use change over the Sudd, would be lower soil moisture and evaporation in DRA compared to the CTL run, whereas the annual evaporation from the Sudd is 1900 mm/yr, the evaporation under drained conditions reduces to 698 mm/yr.

Evaluation of the results from the two simulations (CTL and DRA) has shown that, draining the Sudd can have a large and significant impact on the microclimate over the Sudd. The significance of a changed parameter is evaluated against the inter-annual variability. The reduced evaporation in the drained scenario causes a reduction of the screen level relative humidity by 30 to 40 %, and an increase of the temperature by 4 to 6 °C during the dry season (Fig. 7.3 and 7.4, respectively). The temperature rise is caused by the enhanced sensible heat on the dried wetland. The microclimate impact is confined to the Sudd area and expands somewhat into the west and southwest direction influenced by the wind pattern in the dry season. During the wet season the impact of the Sudd on atmospheric fluxes is not feasible, because the surrounding area is saturated by the rain. However, a slight decrease of relative humidity (<10%) is detected over the Sudd in the wet season.

The simulated results show that the impact of the Sudd on the regional water cycle (atmospheric moisture fluxes, precipitation, evaporation, runoff and sub-surface storage) is negligible and insignificant compared to the inter-annual variability (Fig. 7.7). However, this is not unanticipated with regard to the Sudd evaporated volume relative to the regional hydrologic budget terms. The Sudd evaporated volume is a small fraction relative to the oceanic moisture that is transported laterally through the atmosphere into the basin (less than 1% of the atmospheric moisture in the Nile Basin). However, simulations reveal that the drained Sudd may create climatic perturbations, which generates slightly higher rainfall in the Nile Basin than the present situation. This is most likely attributed to an enhanced convective activity occurred over the dried wetlands with increased atmospheric instability. During the wet season, influence of the Sudd is not noticed. In the dry winter season, when the Sudd may cause largest possible impact on the water cycle, there is no mechanism for rainfall, at least on major parts of the Nile basin (Inter-tropical Convergence Zone ITCZ is retreated, and the upper troposheric tropical easterly jet disappeared).

The implication on the Nile water renewable resources is significant, which constitutes the gain in runoff originating from water that otherwise is evaporated. The net gain of the river flow computed at Malakal (just downstream the Sudd) for draining of the entire Sudd, will be ~ 36 Gm^3/yr. This is slightly less than half the long-term mean of the Nile natural flow at Aswan, and higher than the mean observed losses in the region (29 Gm^3/yr). The mismatch is attributed to bias correction introduced to correct model runoff. The famous Jonglei canal phase one would save about 4 Gm^3/yr, and drain about 30% of the Sudd wetland. Most likely there will be no significant impact on the regional hydrological budget, while the impact on the microclimate over the dried parts might be comparable to the derived results in this study.

The strength of the Sudd experiment presented in this study is that it utilizes a coupled land surface atmosphere model incorporating all possible feedback mechanisms to address a water resources question. The weakness of the experiment is that its time span is limited to only 6 years, making it hard to detect the actual signal caused by land use change from the natural variability in the record. Obviously, one would expect improved results from a longer simulation (e.g., 40 yr) and a finer resolution model (smaller than 50 km grid). Also the hydrological description of river flow processes in a climate model needs to be improved upon. However, the main conclusions derived here are likely to be confirmed rather than discarded, since the hydrological fluxes to and from the Sudd are relatively small compared to the atmospheric fluxes.

Anxiety on the impact of the Jonglei canal has been high with regard to reduction of regional rainfall due to reduced evaporation from the Sudd. This study has proven that this is not the case. However, impact on microclimate, which hasn't been addressed explicitly before can be quite significant. All energy devoted to latent heating will be partitioned into sensible heat, which will increase the local temperature over the drained parts. It is beyond the scope of this study to address all the advantages and disadvantages of draining the Sudd wetland. However, it is pertinent here to propose further investigations to support the attained results: (i)

Cross check the Sudd evaporation estimates derived from remote sensing against ground observation, (ii) Dynamic delineation of the Sudd boundaries, e.g., on monthly time steps, (iii) Longer climate simulation to easily detect actual land use change signal from the natural variability in the record, (iv) Finer model resolution (less than 50 km).

Although the research concentrated on the Nile case study, some results are of generic nature: (i) Confirming the feasibility of using satellite data to estimate evaporation from wetland with scanty ground data, (ii) Verifying that the concept of E_a being proportional to E_w does not hold true for marshlands, (iii) Demonstrating the fact that, the general recycling formula has no prognostic value, although it is useful for qualitative comparisons between different regions, (iv) Demonstrating that river water routing in climate models is a necessity.

Samenvatting

Klimatologie en hydrologie worden steeds meer samengevoegd in één vakgebied, zeker op het niveau van stroomgebieden. Dit komt door de gekoppelde aard van de landoppervlakteprocessen en atmosferische processen. Processen aan het landoppervlak kunnen – b.v. door verdamping – het atmosferische vochttransport beïnvloeden, niet alleen lokaal maar ook op continentale schaal. Er bestaat een positieve terugkoppeling tussen verdamping en regenval in de atmosfeer. Derhalve moet de atmosferische component van de watercyclus in beschouwing genomen worden als men doelt op een complete analyse van de waterbeschikbaarheid op stroomgebiedsniveau.

Het Nijl-stroomgebied karakteriseert zich door snel afnemende waterbeschikbaarheid per hoofd van de bevolking in de benedenstroomse gebieden en hoge verdamping van de bovenstroomse Nijlmoerassen. Al vele jaren zijn er plannen geweest om de afvoer van de Nijl te verhogen door verdamping van deze moerassen te verlagen door middel van een omleidingskanaal (b.v. het Jongleikanaal). De vraag is echter of de verdamping van de bovenstroomse wetlands mede bijdraagt aan de regionale neerslag door vochtrecycling. De hoofdvraag is: hoeveel bedraagt de netto winst in de regionale watercyclus als de wetlands gedraineerd worden?

Het doel van dit onderzoek is om de hoeveelheid vochtrecycling in het Nijl-stroomgebied vast te stellen en te analyseren en om vervolgens te kwantificeren wat de impact van het droogleggen van het Sudd wetland is op de lokale en regionale hydroklimatologie. Regionale klimaatmodellering is het basisgereedschap geweest van het onderzoek.

Allereerst zijn remote sensing technieken ingezet – vanwege de beperkte beschikbaarheid van gegevens over de Nijlmoerassen – om de actuele verdamping van deze uitgestrekte moerassen te schatten. Verdamping vormt een fundamentele randvoorwaarde voor de hydrologische en klimaatmodellen. Het SEBAL algoritme is gebruikt om de actuele verdamping te berekenen met behulp van NOAA-AVHRR beelden. Meer dan 115 satellietbeelden van een gebied van 1000 km x 1000 km (gelegen in de Boven Nijl) zijn verwerkt tot maandelijkse kaarten van verdamping en de biofysische toestand voor 3 jaren met verschillende hydrometeorologische toestanden (1995, 1999 en 2000). De berekende verdamping is geijkt met berekeningen van de waterbalans in 3 van de wetlands: Sudd, Bahr el Ghazal en de Sobat. Een nauwe overeenkomst is verkregen voor de Sudd (2% fout) en de Sobat (5% fout), terwijl de balans niet sluit in het Ghazal gebied (27% fout) door ontoereikende afvoermetingen van de bovenstrooms gelegen deelstroomgebieden. De verdamping van het Sudd wetland is 20% minder, en het gemiddelde oppervlak dat het wetland bestrijkt is 70% groter dan wanneer het beschouwd wordt als een open waterlichaam. Dit weerlegt de aanname die al tientallen jaren lang wordt gebruikt in hydrologische studies van de Sudd. Er is verder diepgaand onderzoek gedaan naar de biofysische toestand zoals afgeleid uit satellietbeelden (oppervlakte albedo, oppervlakte uitstraling, bladoppervlakte-index, ruwheidshoogte, aërodynamische weerstand, oppervlakte weerstand) om de temporele variabiliteit van verdamping uit de wetlands te interpreteren. Er is aangetoond dat de variatie van de atmosferische vraag in combinatie met de jaarlijkse gang van de grondwaterstand – door een duidelijke seizoensafhankelijke regenval en inundatie door de hoogwaters van de Nijl – resulteert in een quasi constante verdampingswaarde uit de Sudd gedurende het jaar.

De ruimtelijk verdeelde verdampingsgegevens van remote sensing boven de Sudd vormt één van de belangrijkste gegevensbronnen voor validatie van een regionaal atmosferisch klimaat model (RACMO) dat het Nijl-stroomgebied dekt. De grens van het model ligt tussen 12°Z en 36°N en tussen 10°E en 54.4°O, met een horizontale resolutie van 50 km. Het model wordt gedreven door ERA-40 gegevens (ECMWF Re-analysis 1957-2001). Het model is zo aangepast, dat het rekening houdt met de complexe hydrologie van de wetlands van de Boven Nijl. De afvoer van de Nijl, die gegenereerd is in de bovenstroomse deelstroomgebieden, voedt het wetland systeem, waar het vervolgens aan verdamping onderhevig is. Er zijn kleine aanpassingen gemaakt in de parametrisatie om orografie (filtering), zonnestraling (aërosols), afvoer (drainage coëfficiënt) en bodemvocht (diepte van de laag) mee te nemen. Dit resulteerde in verbetering van de modelresultaten. Om de modelresulaten te beoordelen zijn tijdreeksen van waarnemingen gebruikt waaronder: straling, neerslag, afvoer en verdampingsgegevens. Het model geeft niet alleen inzicht in de temporele ontwikkeling van de hydroklimatologische parameters van het gebied, maar ook in de interacties tussen landoppervlak en klimaat en de daarin inbegrepen terugkoppelingen. Het model wordt gebruikt om de regionale watercyclus van het Nijl-stroomgebied te beschrijven in de vorm van: atmosferische fluxen, landoppervlak fluxen en terugkoppeling van landoppervlak naar klimaat. De maandelijkse vocht recycling ratio (m.a.w. lokaal gegenereerde neerslag gedeeld door totale neerslag) boven het Nijl-stroomgebied varieert tussen 8 en 14% met een jaargemiddelde van 11%. Dit suggereert dat 89% van de neerslag in het Nijl-stroomgebied afkomstig is van buiten de fysische begrenzing van het stroomgebied. De maandelijkse neerslag-efficiency varieert tussen 12 en 53% met een jaargemiddelde van 28%. De gemiddelde jaarlijkse vochtrecycling ratio in het Nijl-stroomgebied is 11%, wat iets lager is dan in de Amazone maar van dezelfde orde grootte als in het Mississippi-stroomgebied.

De impact van het Sudd-wetland op de hydroklimatologie van de Nijl is bestudeerd door twee scenarios van het regionale klimaatmodel: de huidige klimatologie, en een scenario met een ontwaterde Sudd. De resultaten duiden erop, dat de Sudd over het algemeen een verwaarloosbare invloed heeft op de regionale hydrologische budget termen, vanwege het relatief kleine oppervlak dat het wetland beslaat. De hoeveelheid vochtrecycling door de moerassen is klein vergeleken met de atmosferische fluxen boven het gebied. De afvoertoename in dit extreme scenario zou echter substantieel zijn: ongeveer 36 Gm^3/jr, ca. de helft van de natuurlijke afvoer bij Aswan. De impact op het microklimaat is echter wel groot. De relatieve luchtvochtigheid aan het oppervlak zou dalen met 30 à 40% tijdens het droge seizoen en de temperatuur zou stijgen met 4 tot 6° in het gebied dat de Sudd wetlands momenteel beslaan. De impact tijdens het natte seizoen (juni tot september) is relatief klein.

References

Abtew, W., Obeysekera, J., 1995. Lysimeter Study of Evapotranspiration of Cattails and Comparison of Three Estimation Methods. Transactions of the ASAE, 38(1), 121-129.

Ahmed, M.D. and W.G.M. Bastiaanssen, 2003. Retrieving soil moisture storage in the unsaturated zone from satellite imagery and bi-annual phreatic surface fluctuations. Irrigation and Drainage systems (17) 2: 3-18.

Allen, R.G., L.S. Pereira, D. Raes, and M. Smith, 1998. Crop Evapotranspiration: Guidelines for Computing Crop Water Requirements. United Nations FAO, Irrigation and Drainage Paper 56. Rome, Italy. 300 p.

Allen, R.G., W.G.M. Bastiaanssen, M. Tasumi and A. Morse, 2001. Evapotranspiration on the watershed scale using the SEBAL model and Landsat images, ASAE Meeting Presentation, paper number 01-2224, Sacramento, California, USA, July 30-August 1, 2001.

Allen, R.G., A. Morse, M. Tasumi, R. Trezza, W.G.M. Bastiaanssen, J.L. Wright and W. Kramber, 2002. Evapotranspiration from a Satellite-BASED Surface energy balance for the Snake Plain Aquifer in Idaho, Proc. National USCID conference, San Luis Obispo, California, July 8-10, 2002: 10 pp.

Ashfaque, A., 1999. Estimating lake evaporation using meteorological data and remote sensing. MSc thesis. ITC, Enschede, The Netherlands.

Bastiaanssen, W.G.M., Hoekman, D.H., Roebeling, R.A., 1994. A methodology for the assessment of surface resistance and soil water storage variability at mesoscale based on remote sensing measurements. IAHS special publication no. 2, pp. 66.

Bastiaanssen, W.G.M., Pelgrum, H., Droogers, P., de Bruin, H.A.R. and Menenti, M., 1997. Area-average estimates of evaporation, wetness indicators and top soil moisture during two golden days in EFEDA, Agr. and Forest Met. 87: 119-137.

Bastiaanssen, W.G.M., M. Menenti, R.A. Feddes and A.A.M. Holtslag, 1998a. The Surface Energy Balance Algorithm for Land (SEBAL): Part 1 formulation, J. of Hydr. 212-213: 198-212.

Bastiaanssen, W.G.M., H. Pelgrum, J. Wang, Y. Ma, J. Moreno, G.J. Roerink and T. van der Wal, 1998b. The Surface Energy Balance Algorithm for Land (SEBAL): Part 2 validation, J. of Hydr. 212-213: 213-229.

Bastiaanssen, W.G.M., Bandara, K.M.P.S., 2001. Evaporative depletion assessments for irrigated watersheds in Sri Lanka, Irrigation Science, 21(1), 1-15.

Bastiaanssen, W.G.M., M. Ud-din-Ahmed and Y. Chemin, 2002. Satellite surveillance of evaporative depletion across the Indus Basin, Water Resources Research, vol. 38, no. 12: 1273-1282.

Bastiaanssen, W.G.M. and L. Chandrapala, 2003. Water balance variability across Sri Lanka for assessing agricultural and environmental water use, Agricultural Water Management 58(2): 171-192.

Bauer, P., Brunner, P., Kinzelbach, W., 2002. Quantifying the Net Exchange of Water Between Land and Atmosphere in the Okavango Delta, Botswana, Conf. Proceedings Model Care 2002, 17-20 June 2002, Prag, Czech Republic.

Beljaars, A.C.M., Viterbo, P., Miller, M.J. and Betts, A.K., 1996. The anomalous rainfall over the USA during July 1993: Sensitivity to land surface parameterization and soil moisture anomalies; Mon.Wea.Rev. 124, 362-383.

Benton, G.S., Blackburn, R.T., and Snead V.W., 1950. 'The role of the atmosphere in the hydrological cycle', Trans. Am. Geophys. Union 31, 61-73.

Berger, D.L., Johnson, M.J., Tumbusch, M.L., Mackay, J., 2001. Estimates of Evapotranspiration from the Ruby Lake National Wildlife Refuge Area, Ruby Valley, Northeastern Nevada, May 1999-October 2000. Water-Resources Investigations Report 01-4234, U.S. Geological Survey, U.S. Fish & Wildlife Service.

Berglund, E. R., Mace Jr., A. C., 1972. Seasonal albedo variation of black spruce and sphagnumsedge bog cover types. J. Applied Meteorology 11, 806-812.

Betts, A.K., J.H. Ball, A.C.M. Beljaars, M.J. Miller and P. Viterbo, 1996. The land surface-atmosphere interaction: A review based on observational and global modeling perspectives. J. Geophys. Res., 101D, 7209-7225.

Betts, A.K., J.H. Ball and P. Viterbo, 1999. Basin-scale water and energy budgets for the Mississippi from the ECMWF reanalysis. J. Geophys. Res., 104D, 19,293-19,306.

Bonan, G. B., 1995. Sensitivity of a GCM simulation to inclusion of inland water surfaces. Journal of Climate, 8, 2691-2704.

Boni, G., D. Enthekabi and F. Castelli, 2001. Land data assimilation with satellite measurements for the estimation of surface energy balance components and surface control on evaporation, Water Resources Research, vol. 37, no. 6: 1713-1722

Bosilovich, M.G., and Schubert, S.D., 2001. 'Precipitation recycling in the GEOS-1 data assimilation system over the central United States', J. Hydromet. 2, 26 – 35.

Bosilovich, M.G. and Schubert, S.D., 2002. Water vapor tracers as diagnostics of the regional hydrologic cycle. J. Hydromet. 3:149-165.

Bosilovich, M. G., Y. Sud, S. D. Schubert, and G. K. Walker, 2002. GEWEX CSE sources of precipitation using GCM water vapor tracers, in Global Energy and Water Cycle Experiment NEWS, 3, 6-7.

Bowers, S.A. and R.J. Hanks, 1965. Reflection of radiant energy from soils, Soil Sci. 100(2): 130-138.

Brubaker, K.L., Entekhabi, D. and Eagleson, P.S.: 1993. 'Estimation of continental precipitation recycling', J. Climate, 6, 1077-1089.

Brude, G.I. and Zangvil, A.: 2001. 'The Estimation of Regional Precipitation Recycling. Part I: Review of Recycling Models', J. Climate, 14:2509-2527.

Bruin, de, H.A.R. and Holtslag, A.A.M.: 1989. Evaporation and weather: Interactions with the planetary boundary layer, in (ed.) J. C. Hooghart, Evaporation and Weather, TNO Committee on Hydrological Research, Proc. and Information no. 39: 63-83.

Brutsaert, W.H.: 1982. Evaporation Into the Atmosphere, D. Reidel, Norwell, Mass., p. 199.

Brutsaert, W.H., M. Sugita, Regional surface fluxes from satellite-derived surface temperatures (AVHRR) and radiosonde profiles, 1992. Boundary-Layer Meteorology, 58, pp. 355-366.

Budyko, M.I.: 1974. Climate and Life, Academic Press, p. 508.

Buettner, K.J.K and C.D. Kern 1965. The determination of infrared emissivities of terrestrial surfaces, J. of Geophysical Research, 70, 1329-1337.

Burba, G.G., Verma, S.B., Kim, J., 1999. Surface energy fluxes of Phragmites australis in a prairie wetland. Agriculture and Forest Meteorology, 94 (1), 31-51.

Butcher, A. D., 1938. The Sadd hydraulics. Ministry of Public Works, Cairo

Camberlin, P., 1997. Rainfall anomalies in the Source Region of the Nile and their connection with the Indian Summer Monsoon, J. of Climate, 10, 1380-1392.

Campbell, D. I., Williamson, J.L. 1997. Evapotranspiration from a raised peat bog. J. of Hydrology, 193, 142-160.

Castelli, F., D. Enthekabi and E. Caporali, 1999. Estimation of surface heat flux and an index of soil moisture using adjoint-state surface energy balance, Water Resources Research, 35 (10): 3115-3125.

Chan Siu-On and Peter S. Eagleson, 1980. Water balance studies of the Bahr el Ghazal swamp. Dept. Civ. Engng, Mass. Inst. Tech., Report no. 261

Choudhury, B.J., Idso, S.B., Reginato, R.J., 1987. Analysis of an empirical model for soil heat flux under a growing wheat crop for estimating evaporation by an infrared-temperature based energy balance equation, Agriculture and Forest Meteorology, 39, 283-297.

Choudhury, B.J., 1989. Estimating evaporation and carbon assimilation using infrared temperature data: Vistas in modeling, ed. G. Asrar. Theory and applications in optical remote sensing, John Wiley, New York: 628-690

Conway, D., and M. Hulme, 1996. The impacts of climate variability and climate change in the Nile Basin on future water resources in Egypt, J. Water Resources Development 12(3), 277-296.

Denny, P., 1984. Permanent swamp vegetation of the Upper Nile. Hydrobiologia 110, 79-90

Denny, P., 1991. Africa. In M. Finlayson and M. Moser (Eds), Wetlands. International Waterfowl and Wetlands Research Bureau. Facts on File, Oxford, UK, pp. 115-148

Denny, P., 1993. Wetlands of Africa. In Whigham D.F. et al. (Eds). Wetlands of the World I. Kluwer Academic Publishers, The Netherlands, pp. 1-128

Dolan, T.J., Hermann, A.J., Bayley, S.E., Zoltek, J., 1984. Evapotranspiration of a Florida, USA, freshwater wetland, J. Hydrology, 74, 355-371.

Dolman, A. J., M. Hall, M. L. Kavvas, T. Oki, and J. W. Pomeroy Eds., Soil-Vegetation-Atmosphere Transfer Schemes and Large-Scale hydrological Models (Proceedings of an international symposium (Symposium S5) held during the Sixth IAHS Scientific Assembly at Maastricht, The Netherlands), IAHS Publ. no. 270, July 2001

Driedonks, A.G.M., 1982. Models and observations of the growth of the atmospheric boundary layer, Boundary Layer Meteorology 23: 283-306.

Eagleson, P.S.: 1986. 'The Emergence of Global-Scale Hydrology', Water Resour. Res. 22(9), 6s–14s.

Eisenlohr, W.S., 1966. Water loss from a natural pond through transpiration by hydrophytes, Water Resources Research 2(3), 443-453.

Ek, M. and L. Mahrt, 1994. Daytime Evolution of Relative Humidity at the Boundary-Layer Top. Mon. Wea. Rev., 122, 2709-2721.

Ek, M.B.; Holtslag A.A.M., 2004. Influence of Soil Moisture on Boundary Layer Cloud DEvelopment. Journal of Hydrometeorology 5, 86-99.

El-Hemry, I. I. and Eagleson, P. S., 1980. Water balance estimates of the Machar Marches. Dept. Civ. Engng, Mass. Inst. Tech., Report no. 260

Eltahir, E A. B.,1989. A Feedback Mechanism in Annual Rainfall in Central Sudan, J. of Hydrology, 110, 323-334.

Eltahir, E.A.B.: 1998. `A Soil Moisture-Rainfall Feedback Mechanism, 1. Theory and Observations', Water Resour. Res. 34(4), 765-776.

Eltahir, E A. B. and R. L. Bras, 1994. Precipitation Recycling in the Amazon Basin, Quarterly Journal of the Royal Meteorological Society, 120, 861-880.

Eltahir, E.A.B., and Bras, R.L.: 1996, `Precipitation Recycling', Reviews of Geophysics, 34 (3), 367-379.

El-Tom, M. A. ,1975. The rains of the Sudan, mechanism and distribution, 48 pp., University of Khartoum, Khartoum.

Entekhabi, D., G.R. Asrar, A.K. Betts, K.J. Beven, R.L. Bras, C.J. Duffy, T. Dunne, R.D. Koster, D.P. Lettenmaier, D.B. McLaughlin, W.J. Shuttleworth, M.Th. van Genuchten, M.Y. Wei and E.F. Wood, 1999. An agenda for land surface hydrology research and a call for the second international hydrological decade, Bulletin of the American Meteorological Society, vol. 80, no. 10: 2043-2058.

Farah, H.O., 2001, Estimation of regional evaporation under different weather condition from satellite and meteorological data: a case study in the Naivasha basin, Kenya, Ph.D. thesis, pp. 170, Wageningen University, Wageningen.

Farah, H.O. and W.G.M. Bastiaanssen, 2001. Impact of spatial variations of land surface parameters on regional evaporation: a case study with remote sensing data, Hydr. Processes, vol. 15 (9): 1585-1607

Farmer, G., 1988. Seasonal forecasting of the Kenya coast Short Rains, 1901-84, J. of Climatology, 8, 489-497.

Findell, Kirsten L. and Elfatih A.B. Eltahir, 2003. Atmospheric Controls on Soil Moisture-Boundary Layer Interactions; Part II: Feedbacks Within the Continental United States. The Journal of Hydrometeorology, Vol. 4, No. 3, pp 570-583.

Garratt, J. R., 1993. Sensitivity of climate simulations to land surface and atmospheric boundary-layer treatment-A review, Journal of Climate, 6(3), 419-449.

Gash, J.H.C., Kabat, P., Monteny, B.A., Amadou, M., Bessemoulin, P., Billing, H., Blyth, E.M., deBruin, H.A.R., Elbers, J.A., Friborg, T., Harrison, G., Holwill, C.J., Lloyd, C.R., Lhomme, J.-P., Moncrieff, J.B., Puech, D., Soegaard, H., Taupin, J.D., Tuzet, A. and Verhoef, A. 1997. The variability of evaporation during the HAPEX-Sahel Intensive Observation Period. J. Hydrology 188-189, 385-399.

Gavin, H., Agnew, C.T., 2000. Estimating Evaporation and surface resistance from a wet grassland, Physics and Chemistry of the Earth, 25(7-8), 599-603.

Gilman, K. 1994. Water balance of wetland areas, Conf. on "The balance of water - present and future", AGMET Gp. (Ireland) & Agric. Gp. of Roy. Meteorol. Soc. (UK), Dublin, 7-9 Sep 1994, 123-142

Giorgi, F., Mearns, L.O., Shields, C. and Mayer, L.: 1996. `A regional model study of the importance of local versus remote controls of the 1988 drought and the 1993 flood over the central United States', J. Climate, 9, 1150-1161.

Giorgi, F. and L. O. Mearns, 1999. Introduction to special section: Regional climate modeling revisited, J. of Geophysical Research, 104(D6), 6335-6352.

Griend, van der, A.A. and M. Owe., 1993. on the relationship between thermal emissivity and the normalized difference vegetation index of natural surfaces, Int. J.Of Rem. Sens., 14(6):1,119-131

Hanan, N.P., Prince, S.D., 1997. Stomatal conductance of West Central Supersite vegetation in HAPEX-Sahel: measurements and empirical models. J of Hydrology, 188-189, 536-562.

Hemakumara, H.M., L. Chandrapala and A. Moene, 2003. Evapotranspiration fluxes over mixed vegetation areas measured from large aperture scintillometer, Agr. Water Management 58: 109-122

Herman, A., V. Kumar, P. Arkin, and J. Kousky, 1997. Objectively determined 10-day African rainfall estimates created for famine early warning systems, International Journal of Remote Sensing, 18(10), 2147-2159. the data available at: http://www.cpc.ncep.noaa.gov/products/fews/data.html.

Howell, P.P., Lock, J.M., Cobb, S.M. (Eds) 1988. The Jonglei canal: Impact and Opportunity. Cambridge University Press, U.K, pp. 536.

Hurst, H. E. and Black, R. P., 1931. General description of the Basin, Meteorology, Topography of the White Nile Basin. The Nile Basin, vol. I, Government Press, Cairo.

Hurst, H. E., and Philips, P., 1938. The hydrology of the lake plateau and Baher el Jebel. The Nile Basin, vol. V, Government Press, Cairo.

Idso, S.B., Jackson, R.D., Ehler, W.L. and Mitchell, S.T.: 1969. A method for determination of infrared emittance of leaves, Ecology 40: 899-902.

Idso, S.B. and Anderson, M.G. 1988. A comparison of two recent studies of transpirational water loss from emergent aquatic macrophytes. Aquatic Biology, 31, 191-195.

Jackson, T.J., D.M. Le Vine, A.Y. Hsu, A. Oldak, P.J. Starks, C.T. Swift, J. Isham and M. Haken, 1999. Soil moisture mapping at regional scales using microwave radiometry: The Southern Great Plains Hydrology Experiment, IEEE Trans. Geosci. Rem. Sens.

Jacobs, C.M.J. 1994. Direct impact of atmospheric CO2 enrichment on regional transpiration, Ph.D. thesis, Department of Meteorology, Wageningen University, ISBN 90-5485-250-X: pp. 179.

Jacobs, J.M., Mergelsberg, S. L., Lopera, A.F., Myers, D.A. 2002. Evapotranspiration from a wet prairie wetland under drought conditions: Paynes prairie preserve, Florida, USA. WELANDS, 22(2), 374–385.

Jarvis, P.G. 1976. The interpretation of the variations in leaf water potential and stomatal conductance found in canopies in the field. Philosophical Transactions of the Royal Society of London, (B) 273, 593-610.

Jensen, M.E., (Ed.). 1980. Design and Operation of Farm Irrigation Systems. ASAE Monograph No. 3. Amer. Soc. Agric. Engr. St. Joseph, MI. pp. 829.

Jhorar, R.K., W.G.M. Bastiaanssen, R.A. Feddes and J.C. van Dam, 2002. Inversely estimating soil hydraulic functions using evapotranspiration fluxes, J. of Hydr. 258: 198-213

Jonglei investigation team JIT, 1954. The Equatorial Nile Project and its effects in the Anglo-Egyptian Sudan. Report of the Jonglei Investigation Team, Sudan Government, Khartoum

José, J.S., Meirelles, M.L., Bracho, R., Nikonova, N. 2001. A comparative analysis of the flooding and fire effects on the energy exchange in a wetland community (Morichal) of the Orinoco Llanos. J. of Hydrology, 242 (3-4), 228-254.

Kim, J., Verma, S. 1996. Surface exchange of water vapor between an open sphagnum fen and the atmosphere. Boundary-Layer Meteorology, 79, 243-264.

Kite, G.W. 1998. Land surface parameterization of GCM's and macroscale hydrological models, J. of the American Water Resources Association, 34(6), 1247-1254.

Koch, M. S., Rawlik, P. S. 1993. Transpiration and stomatal conductance of two wetland macrophytes (Cladium jamaicense and Typha domingensis) in the subtropical Everglades. American Journal of Botany, 80(10),1146-1154.

Koerselman, W., Beltman, B. 1988. Evapotranspiration from fens in relation to Penman's potential free water evaporation (E_0) and pan evaporation, Aquatic Botany. 31, 307-320.

Koster, R., J. Jouzel, R. Suozzo, G. Russell, W. Broecker, D. Rind, and P. Eagleson 1986. Global sources of local precipitation as determined by the NASA/GISS GCM. Geophys. Res. Lett., 13, 121-124.

Koster, R.D., Dirmeyer, P.A., Hahmann, A.N., Ijpelaar, R., Tyahla, L., Cox, P., and Suarez, M. J.: 2002. `Comparing the Degree of Land-Atmosphere Interaction in Four Atmospheric General Circulation Models', J. Hydromet. 3, 363-375.

Kustas, W.P., and Daughtry, C.S.T.: 1990. `Estimation of the soil heat flux/net radiation ratio from multispectral data', Agric. For. Meteorol. 49, 205-223.

Kustas, W.P. and J.M. Norman. 1996. Use of remote sensing for evapotranspiration monitoring over land surfaces. Hydrol. Science J. 41(4):495-515

Lafleur, P. M. , W. R. Rouse 1988. The influence of surface cover and climate on energy partitioning and evaporation in a subarctic wetland, Boundary-Layer Meteorology, 44, 327-347.

Lean, J., anl P. R. Rowntree 1997. understaNding thm sensitIvity of(a GCM samulation of Amazonian dMforestation to \he specification of vegmtation ind soil(charactmristics& J. Clieate, 10, 1216-1:35.

Lenderink, G., B. van den Hurk, E. van Meigaard, A. van Ulden, and H. Cuijpers 2003. Simulation of present-day climate in RACMO2: first results and model developments, technical report, pp. 24, KNMI, De Bilt, The Netherlands, 23 July.

Leuning, R., Kelliher, F.M., De Pury, D.G., Schultze, E.D. 1995. Leaf nitrogen, photosynthesis, conductance and transpiration scaling from leaves to canopies, Plant Cell Env., 18, 1183-1200.

Linacre, E.T., Hicks, B.B., Sainty, G.R., Grauze, G. 1970. The evaporation from a swamp. Agric. Meteorology, 7, 375-386.

Lott, R.B., Hunt, R.J. 2001. Estimating evapotranspiration in natural and constructed wetlands. Wetlands, 21,614-628.

Mason, I.M., Harris, A.R., Moody, J.N., Birkett, C.M., Cudlip, W. and Vlachogiannis. Monitoring wetland hydrology by remote sensing: A case study of the Sudd using infra-red imagery and radar altimetr. Proc. of the Central Symp. of the 'International Space Year Conf.', Munich, ESA SP-341, pp79-84, 1992.

Masson, V., J.L. Champeaux, F. Chauvin, C. Mériguet and R. Lacaze, 2003. A global database of land surface parameters at 1km resolution in meteorological and climate models, Journal of Climate, 16, 1261-1282. The data available at: http://www.cnrm.meteo.fr/gmme/PROJETS/ECOCLIMAP/page_ecoclimap.htm.

McNaughton, K.G., Spriggs, T.W. (1986), A mixed model for regional evaporation, Boundary Layer Meteorology, 34, 243-262.

Migahid, A.M. 1948. Report on a Botanical Excursion to the Sudd Region. Fouad I University Press, Cairo, Egypt.

Mohamed, Y. A., W. G. M. Bastiaanssen, H. H. G. Savenije 2004. Spatial variability of evaporation and moisture storage in the swamps of the upper Nile studied by remote sensing techniques, J. of Hydrology, 289, 145-164.

Molion, L.C.B.: 1975. A climatonomic study of the energy and moisture fluxes of the Amazonas Basin with considerations of deforestation effects, Ph.D. thesis, University of Wisconsin, Madison.

Monteith, J.L., Unsworth, M.H., 1990. Principles of environmental physics. Edward Arnold, London, pp. 291.

Moran, S.M. and R.D. Jackson. 1991. Assessing the spatial distribution of evaporation using remotely sensed inputs. J. Environ. Qual. 20:725-737

Morse, A., M. Tasumi, R.G. Allen, W.J. Kramber. 2000. Application of the SEBAL Methodology for Estimating Consumptive Use of Water and Streamflow Depletion in the Bear River Basin of Idaho through Remote Sensing, Final Report, by Idaho Department of Water Resources and University of Idaho. 107 pages

Morton, F.I. 1983. Operational estimates of lake evaporation. J. of Hydrology 66, 77-100.

Nicholson, S.E. 1996. A review of climate dynamics and climate variability in eastern Africa, in The limnology, climatology and paleoclimatology of the East African lakes, edited by T. C. Johnson and E. Odada, pp. 25-56, Gordon and Breach, Amsterdam.

Nieveen, J.P. 1999. Eddy covariance and scintillation measurements of atmospheric exchange processes over different types of vegation, Ph.D. thesis, Department of Meteorology, Wageningen University, ISBN 90-5808-028-5, pp. 121.

Okidi, C.O. 1990. History of the Nile and Lake Victoria Basins through Treaties, in The Nile: Resource evaluation, resource management, hydropolitics and legal issues, edited by P. P.

Pan, Z., E. Takle, M. Segal, and R. Arritt, 1999. Simulation of potential impacts of man-made land use changes on U.S. summer climate under various synoptic regimes. J. Geophy. Res. 104, 6515-6528.

Penfound, W.T., Earle, T.T. 1948. The biology of the water hyacinth. Ecol. Monogr. 18, 449-72.

Penman H L. 1948. Natural evaporation from open water, bare soil and grass. Proc. R. Soc. London Ser. A 193, 120-145.

Penman, H. L. 1963. Vegetation and hydrology. Tech. Comm. No. 53, Commonwealth Bureau of Soils, Harpenden, England. pp. 125.

PJTC 1961. First Annual Report of the Permanent Joint Technical Commission for the Nile Waters, pp. 138, Government Printing Press, Khartoum.

Pielke, R. A., Walko, R. L., Steyaert, L. T., Vidale, P. L., Liston, G. E., Lyons, W. A., & Chase, T. N., 1999. The Influence of Anthropogenic Landscape Changes on Weather in South Florida. Am. Meteorol. Soc.(127), 1663-72.

Perrier, A. 1982. Land surface processes: vegetation. In: (Eds) Eagleson, P.S., Land surface processes in atmospheric general circulation models, Cambridge University Press, Cambridge, pp. 395-448.

Priestley, C.H.B., Taylor, R.J. 1972. On the assessment of surface heat flux and evaporation using large scale parameters. Monthly Weather Review, 100, 81-92.

Raupach, M.R. 1994. Simplified expressions for vegetation roughness length and zero-plane discplacement as functions of canopy height and area index. Boundary Layer Meteorology, 71, 211-216.

Ridder, De K., 1999. The Impact of Surface Evaporative Fraction on Boundary Layer Equivalent Potential Temperature. Phys. Chem. Earth (B), vol. 24, No. 6, pp. 615-618.

Rijks, D.A., 1969. Evaporation from a papyrus swamp, Quart. J. Roy. Meteorol. Soc., 95, 643-649.

Rowntree, R.R., 1988. Review of General Circulation Models for predicting the effects of vegetation change, in (eds). E.R.C. Reynolds and F.B. Thompson, Forests, Climate and Hydrology: Regional Impacts, United Nations University, Oxford: 162-196.

Rudolf, B., Tobias Fuchs, Udo Schneider, and Anja Meyer-Christoffer, 2003. Introduction of the Global Precipitation limatology Centre (GPCC), Global Precipitation Climatology Centre, Deutscher Wetterdienst, Offenbach a.M., Germany. The data available at: http://www.dwd.de/en/FundE/Klima/KLIS/int/GPCC/.

Savenije, H.H.G., 1995. New definitions for moisture recycling and the relation with land-use changes in the Sahel. J. of Hydr. 167:57-78

Savenije, H.H.G., 1996a. The runoff coefficient as the Key to Moisture Recycling. J. of Hydr., 176:219-225, Elsevier, Amsterdam, the Netherlands

Savenije, H.H.G., 1996b. Does Moisture Feedback Affect Rainfall Significantly?. Phys. Chem. Earth, Vol. 20, No. 5-6, pp. 507-513

Savenije, H.H.G., 1997. Determination of evaporation from a catchment water balance at a monthly time scale. Hydrology and Earth System Sciences, Vol. 1, pp. 93-100, EGS, Katlenburg-Lindau, Germany

Schär, Ch., D. Lüthi, U. Beyerle, and E. Heise, 1999. The Soil-Precipitation Feedback: A Process Study with a Regional Climate Model, J. of Climate, 12(3), 722-741.

Schuurmans, J.M., P.A. Troch, A.A. Veldhuizen, W.G.M. Bastiaanssen and M.F.P. Bierkens, 2003. Assimilation of remotely sensed latent heat flux in a distributed hydrological model. Advances in Water Resources 26(2): 151-159

Scott, C.A., W. G. M. Bastiaanssen, and M. D. Ahmad, 2003. Mapping root zone soil moisture using remotely sensed optical imagery, ASCE Irrigation and Drainage Engineering, 129(5), 326-335.

Sellers, P.J., D.A. Randall, G.J. Collatz, J.A. Berry, C.B. Field, D.A. Dazlich, C. Zhang, G.D. Collelo and I. Bounoua, 1996. A revised land surface parameterization (SIB2) for atmospheric GCMs-part 1 – model formulation, J. of Climate, vol. 9: 676-705

Seneviratne, S. I., P. Viterbo, D. Lüthi and C. Schär, 2004. Inferring changes in terrestrial water storage using ERA-40 reanalysis data: The Mississippi River basin. J. Climate,17,2039-2057

Shahin, M., 1985. Hydrology of the Nile Basin, pp. 575, Elsevier, Amsterdam.

Shuttleworth, W.J., 1988. Macrohydrology – the new challenge for process hydrology, J. of Hydr. 100: 31-56.

Smith, M., 1993. CLIMWAT for CROPWAT - A climatic database for irrigation planning and management, FAO Irrigation and Drainage Paper No 49, pp. 113, FAO, Rome.

Smith, Eric, A.Y. Hsu, W.L. Crosson, R.T. Field, L.J. Fritschen, R.J. Gurney, E.T. Kanemasu, W.P. Kustas, D. Nie, W.J. Shuttleworth, J.B. Stewart, S.B. Verma, H.L. Weaver, and M.L. Wesely, 1992. Area-averaged surface fluxes and their time- space variability over the FIFE experimental domain. Journal of Geophysical Research, 97(D17), 18,599-18,622

Small, E.E. and Kurc, S., 2001. The Influence of Soil Moisture on the Surface Energy Balance in Semiarid Environments, technical report, Water Resources Research Institute, New Mexico.

Souch, C., Grimmond, C. S. B., Wolfe, C. P., 1998. Evapotranspiration rates from wetlands with different disturbance histories:Indiana Dunes National Lakeshore. Wetlands 18(2), 216-229.

Stewart, J.B., 1988. Modelling surface conductance of pine forest. Agriculture Forest Meteorology, 43,19-35.

Stewart, J.B. and S.B. Verma, 1992. Comparison of surface fluxes and conductances at two contrasting sites within the FIFE area. J. Geophys. Res. 97 (D17), 18,263-18,628.

Sutcliffe,J. V. and Y. P. Parks, 1999. The Hydrology of the Nile, IAHS Special Publication no. 5, IAHS Press, Institute of Hydrology, Wallingford, Oxfordshire OX10 8BB, UK

Sun, L., F. H. M. Semazzi, F. Giorgi, and L. J. Ogallo, 1999a. Application of the NCAR Regional Climate model to Eastern Africa. Part 1: Simulation of the short rains of 1988, J. Geophys. Res., 104, 6529-6548.

Sun, L. , Semazzi, F. H. M., F. Giorgi, and L. A. Ogallo, 1999b. Application of the NCAR Regional Climate model to Eastern Africa. Part II: Simulation of interannual variability of short rains, J. Geophys. Res., 104, 6549-6562

Taylor, C. M., E. F. Lambin, N. Stephenne, R. J. Harding and R. L. H. Essery 2002. The influence of land use change on climate in the Sahel Journal of Climate 15 3615-3629

Timmer, C.E., Weldon, L.W. 1967. Evapotranspiration and pollution of water by water hyacinth. Hyacinth Control Journal, 6,34-37.

Travaglia C., J. Kapetsky and Mrs. G. Righini, 1995. "Monitoring wetlands for fisheries by NOAA AVHRR LAC thermal data". FAO/SDRN, Rome, Italy

Trenberth, K. E., 1999. Atmospheric moisture recycling: Role of advection and local evaporation. J. Climate, 12, 1368-1381

Todd, M. C., C. Kidd, D. Kniveton, and T. J. Bellerby, 2001. A Combined Satellite Infrared and Passive Microwave Technique for Estimation of Small Scale Rainfall, J. of Atmospheric and Oceanic Technology, 18, 742-755.

Tucker, C.J. 1979. Red and photographic infrared linear combinations for monitoring vegetation. Remote Sensing of the Environment, v. 8, p.127-150

van den Hurk, B. J. J. M., P. Viterbo, A. C. M. Beljaars, and A. K. Betts 2000, Offline validation of the ERA40 surface scheme, ECMWF TechMemo 295, Reading, U.K.

van der Weert, R., Kamerling, G.E., 1974. Evapotranspiration of water hyacinth (Eichhornia crassipes), J. Hydrology 22, 201-212.

Verhoef, A., McNaughton, K.G., Jacobs, A.F.G. 1997. A parameterization of momentum roughness length and displacement height for a wide range of canopy densities. Hydrology and Earth Systems Sciences 1, 81-91.

Walker, J.P., G.R. Willgoose and J.D. Kalma, 2001. One-dimensional soil moisture profile retrieval by assimilation of near-surface observations: a comparison of retrieval algorithms, Advances in Water Resources 24: 631-650

Watson, R.T. and the Core Writing Team (Eds.) 2001. Third Assessment Report of the Intergovernmental Panel on Climate Change, Working Group II, pp 398, Cambridge University Press, UK.

Wieringa, J. 1986. Roughness-dependent geographical interpolation of surface wind speed averages, Quart. J. Roy. Meteo. Soc. 112, 867-889

Zheng, X. and Eltahir, E.A.B.: 1998. 'A Soil Moisture-Rainfall Feedback Mechanism, 2. Numerical Experiments', Water Resour. Res. 34(4), 777-786.

Zhong, Q. and Y.H. Li. 1988. Satellite observation of surface albedo over the Q_{in}ghai-Xizang plateau region, Adv. In Atm. Sciences 5:57-65

Appendix A: NOAA-AVHRR satellite images

Table A1: Acquisition date of the NOAA-AVHRR scenes in year 1995 used in the calculation

Date of acquisition 1995							
01.	24-Jan	12.	18-Apr	23.	04-Jul	34.	03-Oct
02.	25-Jan	13.	28-Apr	24.	12-Jul	35.	10- Oct
03.	26-Jan	14.	07-May	25.	21-Jul	36.	13- Oct
04.	06-Feb	15.	16-May	26.	09-Aug	37.	28- Oct
05.	10-Feb	16.	24-May	27.	16-Aug	38.	06-Nov
06.	20-Feb	17.	26-May	28.	17-Aug	39.	21-Nov
07.	04-Mar	18.	04-Jun	29.	25-Aug	40.	28-Nov
08.	15- Mar	19.	12-Jun	30.	05-Sep	41.	08-Dec
09.	31- Mar	20.	22-Jun	31.	17-Sep	42.	15-Dec
10.	01-Apr	21.	24-Jun	32.	21-Sep	43.	28-Dec
11.	17-Apr	22.	02-Jul	33.	23-Sep		

Table A2: Acquisition date of the NOAA-AVHRR scenes in year 1999 used in the calculation

Date of acquisition 1999							
01.	03-Jan	10.	07-Apr	19.	06-Jul	28.	05-Oct
02.	17-Jan	11.	17-Apr	20.	15-Jul	29.	13- Oct
03.	26-Jan	12.	26-Apr	21.	25-Jul	30.	31- Oct
04.	06-Feb	13.	05-May	22.	03-Aug	31.	07-Nov
05.	15-Feb	14.	13-May	23.	12-Aug	32.	16-Nov
06.	24-Feb	15.	25-May	24.	22-Aug	33.	25-Nov
07.	06-Mar	16.	02-Jun	25.	05-Sep	34.	05-Dec
08.	14- Mar	17.	11-Jun	26.	15-Sep	35.	13-Dec
09.	24- Mar	18.	26-Jun	27.	25-Sep	36.	21-Dec

Table A3: Acquisition date of the NOAA-AVHRR scenes in year 2000 used in the calculation

Date of acquisition 2000							
01.	06-Jan	11.	12-Apr	21.	30-Jul	31.	25- Oct
02.	17-Jan	12.	27-Apr	22.	10-Aug	32.	09-Nov
03.	26-Jan	13.	07-May	23.	18-Aug	33.	18-Nov
04.	02-Feb	14.	18-May	24.	26-Aug	34.	28-Nov
05.	11-Feb	15.	25-May	25.	27-Aug	35.	06-Dec
06.	28-Feb	16.	04-Jun	26.	06-Sep	36.	15-Dec
07.	07-Mar	17.	18-Jun	27.	14-Sep	37.	23-Dec
08.	16- Mar	18.	27-Jun	28.	22-Sep		
09.	27- Mar	19.	05-Jul	29.	07-Oct		
10.	02-Apr	20.	16-Jul	30.	17- Oct		

Appendix B: Hydrological data in the study area

Table B1: Sudd Basin: Mean record of input data and water balance results based on Sutcliffe and Parks (1999) model.

Month	P mm/month	E mm/month	R_{in} Gm³/month	R_{out} Gm³/month	S Gm³	dS/dt Gm³/month	A Gm²
Jan	2	217	3.89	1.95	19.64	-2.56	19.64
Feb	3	190	3.37	1.63	17.86	-1.77	17.86
Mar	22	202	3.62	1.77	16.61	-1.25	16.61
Apr	59	186	3.60	1.62	16.51	-0.10	16.51
May	101	183	4.00	1.58	17.38	0.87	17.38
Jun	116	159	3.86	1.50	18.72	1.34	18.72
Jul	159	140	4.19	1.51	21.33	2.61	21.33
Aug	160	140	4.65	1.59	24.36	3.03	24.36
Sep	136	150	4.61	1.70	26.59	2.22	26.59
Oct	93	177	4.69	1.94	27.03	0.44	27.03
Nov	17	189	4.47	1.94	25.07	-1.95	25.07
Dec	3	217	4.21	2.06	22.17	-2.91	22.17
Total	871	2150	49.16	20.80		-0.03	
Avg							21.11

Table B2: Baher el Ghazal Sub-basin: Mean record of input data and water balance results based on Sutcliffe and Parks (1999) model.

Month	P mm/month	E mm/month	R_{in} Gm³/month	R_{out} Gm³/month	S Gm³	dS/dt Gm³/month	A Gm²
Jan	0	217	0.05	0.02	7.26	-1.74	7.26
Feb	4	190	0.01	0.03	6.01	-1.26	6.01
Mar	14	202	0.00	0.04	4.94	-1.07	4.94
Apr	49	186	0.01	0.04	4.38	-0.56	4.38
May	110	183	0.16	0.03	4.23	-0.16	4.23
Jun	143	159	0.47	0.02	4.54	0.32	4.54
Jul	175	140	1.06	0.02	5.58	1.04	5.58
Aug	184	140	2.01	0.03	7.54	1.96	7.54
Sep	141	150	3.05	0.02	10.10	2.57	10.10
Oct	69	177	2.92	0.02	11.62	1.52	11.62
Nov	10	189	1.29	0.01	10.88	-0.74	10.88
Dec	1	217	0.31	0.01	9.02	-1.86	9.02
Total	900	2150	11.33	0.31	7.18	0.02	
Avg							7.18

Table B3: Sobat Sub-basin: Input data based on Sutcliffe and Parks (1999).

Month	P mm/month	E mm/month	R_{in} Gm^3/month	R_{out} Gm^3/month
Jan	0	217	0.35	1.02
Feb	2	190	0.23	0.45
Mar	4	202	0.22	0.29
Apr	37	186	0.28	0.25
May	130	183	0.62	0.43
Jun	151	159	1.57	0.86
Jul	214	140	2.65	1.29
Aug	288	140	3.53	1.59
Sep	166	150	4.05	1.77
Oct	92	177	2.75	1.99
Nov	31	189	1.11	1.98
Dec	0	217	0.60	1.76
Total	1115	2150	17.96	13.69

Appendix C: Temporal variability of evaporation and biophysical properties

Table C1: Measured E_a/E_w values of various wetlands spread around the world.

No.	E_a/E_w	Wetland Vegetation	Location	Source and methodology
1	0.7-0.9	Hydrophytes (bulrush, cattail, white top)	Dakota, USA	Eisenlohr (1966), Water balance
2	0.6±0.15	Papayrus	Uganda	Rijks (1969), E_a^1, E_w^2
3	0.7	Typha reeds	Australia	Linacre et al. (1970), E_a^3, E_w by miscellaneous formulae
4	~ 1.5	Water haycinth	Surinam	Van der Weert and Kamerling (1974) E_a measured in water tank, E_w from class A pan.
5	1±0.15	Sphagnum (papill osum and majus), vascular plants	Minnesota, USA	Kim and Verma (1996), E_a^1, E_w^2
6	0.34 – 0.77	Raised peat bog Sphagnum (Empodisma minus)	New Zealand	Campbell and Williamson (1997), E_a^1, E_w^2
7	~ 1.0	Common arrowhead, yellow pond lily, cattail	Indiana, USA	Souch et al., (1998), E_a^3, E_w by P-M
8	0.75-1.0	Phragmites Australis (reed grass), early and peak growth	Nebraska, USA	Burba et al. (1999), E_a^1, E_w^2
9	0.8-1.5	Water hyacinth	Kenya	Ashfaque (1999, pp-A10), E_a^1 E_w by P-M (r_s=0)
10	0.84 – 1.10	Natural wetland grass	England	Gavin and Agnew (2000), E_a from soil moisture balance, E_w by P-M (r_s=0)
11	0.78	Palm, flooded herbaceous (unburned)	Venzuela	José et al. (2001), E_a^1, E_a^3, E_w is from type A tank
Avg.	0.87			
σ	0.26			

E_a^1 = wetland evaporation by Bowen ratio
E_w^2 = Open water evaporation by Penman (1948)
E_a^3 = wetland evaporation by eddy correlation

Table C2: Measured biophysical properties over selected wetlands around the world

No.	I_{NDV} (-)	r_0 (-)	r_s (s/m)	z_{0m} (m)	Λ (-)	Vegetation type	Source
1					0.70	Papayrus, (old papayrus)	Rijks (1969), Uganda
2		0.12-0.16				Sphagnum-sedge bogs	Berglund and Mace (1972), Minnesota, USA
3		0.18	0.0			Water hyacinth	Van der Weert and Kamerling (1974), Surinam
4	1.9-2.5		50-107	0.04		Sedge (Carex paleacea)	Lafluer and Rouse 1988), (Canada, Backshore)
5	1.9-2.5		39-87			Marches: carex paleacea,	Lafleur and Rouse (1988) (Canada, Open water Marches)
6	5.0		24-37	0.41		Woodland (Alnus rugosa, Salix bebbianna, others)	Lafleur and Rouse (1988), (Canada, Woodland)
7	0.4-0.7	0.11-0.17	80-250		0.76 – 1.0	Sphagnum (papill osum and majus), vascular plants	Kim and Verma (1996), Minesota, USA
8	1.25-3.9		150-608	0.091	0.16-0.25	Raised peat bog Sphagnum (Empodisma minus)	Campbell and Williamson (1997), New Zealand
9			0-5		0.67– 0.74	Common arrowhead, yellow pond lily, cattail	Souch et al. (1998), Indiana USA
10		0.17		0.009	0.99	Water hyacinth	Ashfaque (1999, pp-A10)
11	1.2-2.6	0.12-0.16		0.27-0.38	0.75 – 1.0	Phragmites australis (reed grass), early and peak growth	Bruba et al. (1999), Nebraska, USA. (z_{0m} calculated as 0.123h)
12			8-155			Natural wetland grass	Gavin and Agnew (2000), England
13		0.16	25-100	0.07-0.08	0.78	Palm, flooded Herbaceous (unburned)	Jose et al. (2001), Venzuela, (z_{0m} calculated as 0.123h)
14			50		0.73	Maiden cane, mock bisgop's weed, dog fennel	Jacobs et al. (2002), (wet period) Florida, USA
Avg.	2.4	0.16	86	0.16	0.73		
σ	1.45	0.02	97	0.17	0.24		

Table C3: Remote sensing assessment of the temporal variability of the monthly biophysical and meteorological parameters in the Sudd based on 3 year average (1995, 1999, 2000) for 3.86×10^6 ha area

Month	R_n W/m²	$(e_s - e_a)$ kPa	Δ kPa/°C	I_{NDV} (-)	r_0 (-)	z_{0m} m	ε_0 (-)	r_a s/m	r_s s/m	Λ (-)
Jan	121.2	2.71	0.227	0.34	0.19	0.019	0.94	53	349	0.89
Feb	128.5	3.18	0.230	0.22	0.18	0.013	0.92	56	404	0.90
Mar	143.2	3.28	0.240	0.26	0.18	0.015	0.93	53	477	0.83
Apr	136.7	2.81	0.250	0.23	0.20	0.014	0.93	58	305	0.89
May	133.9	1.92	0.233	0.44	0.19	0.025	0.94	61	177	0.94
Jun	124.3	1.42	0.213	0.50	0.20	0.028	0.95	63	123	0.93
Jul	119.5	0.98	0.197	0.53	0.19	0.029	0.95	63	72	0.94
Aug	126.9	0.92	0.190	0.72	0.17	0.042	0.96	61	55	0.95
Sep	130.7	1.09	0.207	0.68	0.18	0.039	0.96	67	75	0.94
Oct	123.5	1.27	0.180	0.59	0.20	0.032	0.96	72	94	0.97
Nov	123.4	2.17	0.207	0.41	0.18	0.023	0.94	65	244	0.93
Dec	117.8	2.45	0.203	0.32	0.17	0.018	0.93	59	308	0.91
Avg.	127.5	2.02	0.21	0.44	0.19	0.025	0.94	60.86	223.6	0.92

Table C4: Monthly evaporation values for a 38×10^6 ha area in the Sudd during 3 years

Month	E_a 1995 mm/day	E_a 1999 mm/day	E_a 2000 mm/day	E_a Avg.	E_a/E_w 1995	E_a/E_w 1999	E_a/E_w 2000	E_a/E_w Avg.
Jan	3.7	4.6	4.6	4.29	0.55	0.55	0.58	0.56
Feb	3.7	5.6	4.3	4.52	0.50	0.64	0.59	0.58
Mar	4.3	4.9	4.4	4.55	0.49	0.59	0.54	0.54
Apr	4.1	6.2	4.9	5.05	0.54	0.65	0.62	0.60
May	4.3	5.9	4.8	4.98	0.67	0.79	0.74	0.73
Jun	4.1	5.3	4.5	4.61	0.74	0.81	0.72	0.76
Jul	4.0	5.2	4.2	4.46	0.82	0.89	0.8	0.84
Aug	4.1	5.3	4.7	4.72	0.84	0.93	0.91	0.89
Sep	4.4	5.5	4.4	4.74	0.84	0.93	0.78	0.85
Oct	4.1	4.9	4.9	4.63	0.75	0.87	0.8	0.81
Nov	3.8	5.5	4.1	4.48	0.63	0.77	0.64	0.68
Dec	3.6	4.8	4.3	4.23	0.57	0.65	0.63	0.62
Avg.	4.0	5.3	4.5	4.61	0.66	0.76	0.70	0.70

Table C5: First order assessment of the groundwater table behavior for the years 1995, 1999 and 2000 (θ soil moisture relative to saturation).

Month	1995		1999		2000	
	θ/θ_{sat}	GWT	θ/θ_{sat}	GWT	θ/θ_{sat}	GWT
Jan	0.73	-0.30	0.82	-0.79	0.87	-0.66
Feb	0.70	-0.46	0.86	-0.99	0.88	-0.72
Mar	0.57	-0.73	0.81	-1.12	0.80	-0.81
Apr	0.68	-0.74	0.89	-1.26	0.87	-0.89
May	0.80	-0.72	0.93	-1.37	0.91	-0.85
Jun	0.84	-0.51	0.91	-1.20	0.80	-0.77
Jul	0.89	-0.25	0.94	-1.11	0.84	-0.34
Aug	0.89	-0.21	0.94	-0.88	0.89	-0.19
Sep	0.93	-0.09	0.95	-0.65	0.78	-0.11
Oct	0.92	0.00	0.95	-0.46	0.93	0.00
Nov	0.86	-0.13	0.90	-0.58	0.85	-0.04
Dec	0.74	-0.28	0.88	-0.63	0.88	-0.17

Appendix D: RACMO model results over the Nile Basin.

Table D1: model results of the Mean annual cycle 1995 to 2000 in mm/day.

Month	Q_{in}	Q_{out}	P	E	R	dS/dt
Jan	5.16	5.62	0.63	0.96	0.11	-0.40
Feb	4.93	5.32	0.62	0.94	0.10	-0.36
Mar	6.29	5.93	1.38	1.11	0.16	0.11
Apr	6.44	5.72	1.73	1.31	0.21	0.20
May	5.88	5.66	1.61	1.44	0.17	0.01
Jun	5.17	4.99	1.39	1.26	0.14	0.03
Jul	5.63	4.47	2.38	1.51	0.28	0.60
Aug	5.70	4.77	2.76	1.77	0.45	0.49
Sep	5.58	5.42	1.84	1.66	0.31	-0.14
Oct	5.51	5.31	1.84	1.50	0.29	0.02
Nov	5.31	5.42	1.32	1.26	0.21	-0.20
Dec	5.21	5.61	0.79	1.03	0.14	-0.36

about the author

Yasir Abbas Mohamed was born in Sennar, Sudan on July 05, 1961. He obtained his B.Sc. (div1) in Civil Engineering in 1985 from the University of Khartoum. Since then he joined the Hydraulic Research Station (HRS) in Wad Meddani of the Ministry of Irrigation of Sudan as an assistant research engineer, where he was involved in irrigation water management research in the Gezira scheme. He obtained his diploma (with distinction) on land and water development from IHE in 1989, and subsequently his MSc (with distinction) in 1990 on the thesis "Simulation and optimization of the Blue Nile double reservoir system". During his MSc study at IHE, he also participated in the research: ground water modeling of the Bergambacht area, The Netherlands. Through a sabbatical leave from HRS, he joined WL | Delft Hydraulics in 1991/92, where he participated in all stages of development of the flood early warning system for the Nile river in Sudan (including: data collection, river surveys, hydrologic and hydrodynamic modeling). In 1993/94 back at HRS, he was involved in water management research in the Gezira scheme, and development of new operating policies for the Blue Nile reservoirs. He lectured the course on Sudan's water resources to the graduates of the University of Gezira, Wad Meddani. During 1995/98 he joined WL | Delft Hydraulics as a river flow modeler and participated in all stages of developing the flood forecasting system for the Indus river and tributaries in Pakistan. In 1998/99, back at HRS he led two teams: one working on the water balance study of the Blue Nile River, and a team working on the hydrologic design of small dams across the Dinder River. In 1999 he worked in the capacity of the Secretary General of the Sudan National Committee on Irrigation and Drainage (SNCID). During 2000/2002 he worked as a senior researcher at the hydraulic lab of the Abu Dhabi Water and Electricity Authority, Abu Dhabi, UAE, where he developed 2D and 3D (Delft 3D) models for water circulation studies in the Abu Dhabi lagoon system. Since July 2001 he is working on his PhD research. The first year was sandwiched with ITC, Enschede, where he prepared evaporation maps of the Sudd area. As part of the literature review he conducted an international electronic discussion forum on moisture recycling in the summer of 2003 (associated with the Dialogue on Water and Climate). He spent 1½ year at the Royal Netherlands Meteorological Institute (KNMI) in De Bilt, where he was involved in the climate modeling part of the PhD study. He has several publications in international journals and conferences.